Studies in Big Data

Volume 28

Series editor

Janusz Kacprzyk, Polish Academy of Sciences, Warsaw, Poland
e-mail: kacprzyk@ibspan.waw.pl

About this Series

The series "Studies in Big Data" (SBD) publishes new developments and advances in the various areas of Big Data- quickly and with a high quality. The intent is to cover the theory, research, development, and applications of Big Data, as embedded in the fields of engineering, computer science, physics, economics and life sciences. The books of the series refer to the analysis and understanding of large, complex, and/or distributed data sets generated from recent digital sources coming from sensors or other physical instruments as well as simulations, crowd sourcing, social networks or other internet transactions, such as emails or video click streams and other. The series contains monographs, lecture notes and edited volumes in Big Data spanning the areas of computational intelligence incl. neural networks, evolutionary computation, soft computing, fuzzy systems, as well as artificial intelligence, data mining, modern statistics and Operations research, as well as self-organizing systems. Of particular value to both the contributors and the readership are the short publication timeframe and the world-wide distribution, which enable both wide and rapid dissemination of research output.

More information about this series at http://www.springer.com/series/11970

Vicenç Torra

Data Privacy: Foundations, New Developments and the Big Data Challenge

 Springer

Vicenç Torra
School of Informatics
University of Skövde
Skövde
Sweden

ISSN 2197-6503   ISSN 2197-6511   (electronic)
Studies in Big Data
ISBN 978-3-319-86141-8   ISBN 978-3-319-57358-8   (eBook)
DOI 10.1007/978-3-319-57358-8

Printed on acid-free paper

This Springer imprint is published by Springer Nature
The registered company is Springer International Publishing AG
The registered company address is: Gewerbestrasse 11, 6330 Cham, Switzerland

To Klara and to my children

Preface

Data privacy is now a hot topic. Big data have increased its importance. Nevertheless, computational methods for data privacy have been studied and developed since the 70s, at least.

I would say that there are three different communities that work on data privacy from a technological perspective. One is the statistical disclosure control (people with a statistical background), another is the privacy-preserving data mining (people that proceed from databases and data mining), and finally the privacy-enhancing technologies (people that proceed from communications and security).

This book tries to give a general perspective of the field of data privacy in a unified way, and cover at least partially some of the problems and solutions studied by these three communities. The goal is not to present all methods and algorithms for all type of problems, nor the latest algorithms and results. I think that this is an almost impossible task. The goal is to give a broad view of the field, present the different approaches (some of them in their basic form), so that the reader can then deepen in their field of interest.

In this way, the book differs from others that focus on only one of the areas of data privacy. For example, we find the following reference books on statistical disclosure control [1–3], privacy-preserving data mining [4] (edited book), [5], and [6] (focusing on computation-driven/cryptographic approaches). We also edited [7], a book that presents the research on privacy in the ARES project. In addition, there are other books that focus on a specific type of data protection approach or type of problem (e.g., association rule hiding [8], data outsourcing [9], synthetic data generators [10], differential privacy [11], databases and microaggregation [12]). A book that gives a broad picture of privacy for big data is [13] but, in contrast to the others, it does not include details on data privacy technologies.

I have been working on data privacy for more than 15 years. My research has mainly focused on topics related to privacy for databases, and disclosure risk measurement. Because of that, the book is biased into data protection mechanisms for databases.

The book is partially based on the lectures I gave at the Universitat Autònoma de Barcelona and at the University of Linköping; and material from this book was used in these courses.

The book is written for last courses of undergraduate studies in computer engineering, statistics, data science and related fields. The book is expected to be self-contained.

Organization

The structure of the book is as follows. Chapter 1 gives an introduction to the field, reviewing the main terminology. Chapter 2 is a brief summary on techniques in machine and statistical learning. We review those tools that are needed later on. Chapter 3 gives a road-map of data protection procedures. It contains a classification of procedures from different perspectives. The chapter includes two sections on specific data protection methods. One related to result-driven approaches for association rules and the other related to methods for tabular data. Chapter 4 focuses on methods for user's privacy. We discuss methods in communication and for information retrieval.

Chapter 5 discusses privacy models and disclosure risk measures. Naturally, we discuss attribute and identity disclosure. The chapter includes a description of methods for data matching and record linkage as they are used for assessment of disclosure risk.

Chapter 6 is about masking methods (data protection methods for respondent and holder privacy). Literature on masking methods is very large. In this chapter we try to describe the major families of methods, and we include some algorithms of these families. We also refer to some alternative methods, but the chapter does not intend to be exhaustive (an impossible task, indeed!). Selection has been based on simplicity of the algorithm, well-knownness, and personal bias.

Chapter 7 is about information loss and data utility. We review the main approaches for evaluating information loss. This chapter tries to give a broad coverage of the alternative ways used in the literature to evaluate masking methods.

The book finishes with two final chapters, one (Chap. 8) on the selection of a masking method based on disclosure risk and information loss, and another (Chap. 9) that concludes the book.

How to Use This book

The book can be used to give a general introduction on data privacy, describing the different approaches in SDC, PPDM, and PETs. For this, the course would use the first and third chapters of the book. If the course is expected to be technical, then we would use most of the material in the book.

Alternatively, it can be used focusing on masking methods. In this case, emphasis would be given to Chap. 6, on the methods, with an overview of Chaps. 5 and 7.

The book can be used focusing on data privacy for big data. It contains a description of data privacy methods for big data. In Chap. 6 (on masking methods) there is for each family of methods a section focusing on big data, and Sect. 6.6, at the end of the chapter, wraps up all these partial discussions. The sections on each family of methods and their use for big data are as follows. Section 6.1.1 is on rank swapping, Sect. 6.1.2 on microaggregation, Sect. 6.1.3 on additive and multi-plicative noise, Sect. 6.1.4 on PRAM, Sect. 6.1.5 on lossy compression, and Sect. 6.4.2 on algorithms for k-anonymity for big data. Then, we also discuss disclosure risk and information loss for big data in the corresponding chapters. That is, Sect. 5.9.3 is on disclosure risk (including a subsection on guidelines and research issues) and Sect. 7.5 is on information loss.

I used parts of this book in courses on data privacy with 6 and 12 h of lectures. The course with 12 h described the main concepts of data privacy, the classification of methods, and a high-level description of disclosure risk and information loss, and a summary of masking methods. A course syllabus for 8 lectures of 2 h follows. The structure maps the chapters of this book.

- L1. Introduction to data privacy.
- L2. Elements of statistical and machine learning.
- L3. Classification of data protection mechanisms.
- L4. Privacy models.
- L5. Masking methods I.
- L6. Masking methods II.
- L7. Information loss. Masking method selection.
- L8. Other privacy protection mechanisms. Result-driven approaches. Methods for tabular data. Methods for user privacy.

The book does not contain exercises. For experimentation, open software as the sdcMicro package [14] in R can be used. I used this in my courses. Another open-source software for data anonymization is ARX [15].

Acknowledgements

I was introduced to this field by Josep Domingo-Ferrer when we met in Tarragona, at the Universitat Rovira i Virgili, at the end of the 1990s. So, my acknowledgement first goes to him.

Second, special thanks go to former (Ph.D.) students and postdocs of my research group with whom we have researched in different areas of data privacy: Aïda Valls, Jordi Nin, Jordi Marés, Jordi Casas, Daniel Abril, David Nettleton, Sergi Martínez, Marc Juarez, Susana Ladra, Javier Jimenez, Julián Salas, Cristina Martínez, Javier Herranz, and Guillermo Navarro-Arribas.

Third, to J. Lane, J. Castro, N. Shahmehri, and the people of the ARES and CASC projects. During the period 2008–2014 most of our research on data privacy was funded by the Spanish CONSOLIDER research project titled ARES. The project gathered most of the research groups in Spain working in the field of privacy and security. CASC was an EU project (2001–2004) on statistical confidentiality. Part of my research described here was funded by these projects. The research of my own described in this book was performed while working at IIIA-CSIC, and lately at SAIL group at the U. of Skövde.

Parts of this manuscript were read and commented by Eva Armengol, Yacine Atif, and Guillermo Navarro. Special thanks go to them.

Last but not least, thanks to my family for their help, and particularly to Klara for the long discussions on privacy-related issues.

Naturally, all errors in the book (except the ones to avoid disclosure) are mine.

Cap de creus and l'Escala   Vicenç Torra
August 2016

References

1. Willenborg, L., de Waal, T.: Elements of Statistical Disclosure Control. Lecture Notes in Statistics. Springer, New York (2001)
2. Duncan, G.T., Elliot, M., Salazar, J.J.: Statistical Confidentiality. Springer, New York (2011)
3. Hundepool, A., Domingo-Ferrer, J., Franconi, L., Giessing, S., Nordholt, E.S., Spicer, K., de Wolf, P.-P.: Statistical Disclosure Control. Wiley, New York (2012)
4. Aggarwal, C.C., Yu, P.S. (eds.): Privacy-Preserving Data Mining: Models and Algorithms. Springer, New York (2008)
5. Fung, B.C.M., Wang, K., Fu, A.W.-C., Yu, P.S.: Introduction to Privacy-Preserving Data Publishing: Concepts and Techniques. CRC Press (2011)
6. Vaidya, J., Clifton, C.W., Zhu, Y.M.: Privacy Preserving Data Mining. Springer, New York (2006)
7. Navarro-Arribas, G., Torra, V. (eds.): Advanced Research in Data Privacy. Springer (2015)
8. Dasseni, E., Verykios, V.S., Elmagarmid, A. K., Bertino, E. Hiding association rules by using confidence and support (2001)
9. Foresti, S.: Preserving Privacy in Data Outsourcing. Springer (2011)
10. Drechsler, J.: Synthetic Datasets for Statistical Disclosure Control: Theory and Implementation. Springer, New York (2011)
11. Li, N., Lyu, M., Su, D., Yang, W.: Differential Privacy: From Theory to Practice, Morgan and Claypool publishers (2016)
12. Domingo-Ferrer, J., Sánchez, D., Soria-Comas, J.: Database Anonymization: Privacy Models, Data Utility, and Microaggregation. Morgan and Claypool publishers (2016)
13. Lane, J., Stodden, V., Bender, S., Nissenbaum, H. Privacy, big data, and the public good. Cambridge University Press (2014)
14. Templ, M.: Statistical disclosure control for microdata using the R-Package sdcMicro. Trans. Data Priv. 1, 67–85 (2008)
15. http://arx.deidentifier.org/. Accessed Jan 2017

Contents

Introduction

1

Neuk ere izan nuen ezkutuko maite bat

K. Uribe, Bitartean heldu eskutik, 2001, p. 46 [1]

Data is nowadays gathered in large amounts by companies and governmental offices. This data is often analyzed using statistical and data mining methods. When these methods are applied within the walls of the company that has gathered the data, the danger of disclosure of sensitive information might be limited. In this case it is mainly an issue of technologies for ensuring security and access control. In contrast, when the analysis has to be performed by third parties, privacy becomes a much more relevant issue. Similar problems arise when other actors not directly related to the data analysis enter into the scene (e.g., software developers who need to develop and test procedures on data they are not allowed to see).

To make matters worst, it is not uncommon the scenario where an analysis does not only require data from a single data source, but from several sources. This is the case of banks tracking fraud detection and hospitals analyzing diseases and treatments. In the first case, data from several banks might help in fraud detection. Similarly, data from different hospitals might help in the process of finding the causes of a bad clinical response to a given treatment, or the causes of a given disease.

Privacy enhancing technologies (PET), Privacy-Preserving Data Mining [2] (PPDM), and Statistical Disclosure Control [3,4] (SDC) are related fields with a similar interest on ensuring data privacy. Their goal is to avoid the disclosure of sensitive or proprietary information to third parties.

Privacy appeared first within the statistical community in relation to the publication of data from census. For example, Dalenius work [5] dates back to 1977. In order to avoid the disclosure of sensitive data from respondents, tools and methodologies were developed. Research mainly focuses on statistical databases and users

© Springer International Publishing AG 2017
V. Torra, *Data Privacy: Foundations, New Developments and the Big Data Challenge*, Studies in Big Data 28, DOI 10.1007/978-3-319-57358-8_1

typically are decision makers and social scientists. Later the computer science community became also interested in this topic involving people working on database and data mining on the one hand, and people working on data communications on the other hand. In databases and data mining, researchers faced problems similar to the ones of the statistical community but in this case issues are related to the exploitation of data from companies. In this area, it is common to use proprietary databases (e.g., data from banks, hospitals, economic transactions) in data mining applications. Accordingly, the focus, the type of data and the data uses are slightly different. First papers appeared in late 1990s (see e.g. [6–9]). In data communications, researchers studied privacy and security with respect to data tranmission (see e.g., Chaum's work [10] published in 1981). Anonymous communication falls in this area. Within these fields, several methods have been proposed for processing and analysing data without compromising privacy, and for releasing data ensuring some levels of data privacy.

Privacy models have been defined to establish when data does not lead to disclosure and, thus, can be released. Measures and indices have been defined for evaluating disclosure risk (that is, to what extent data satisfy the privacy constraints), and for evaluating data utility or information loss (that is, to what extent the protected data is still useful for applications). In addition, tools have been proposed to visualize and compare different approaches for data protection. All these elements will be presented in this book.

In this chapter we discuss why methods for data privacy are used, we underline some links between data privacy and society, and conclude with a review of the terminology related to data privacy. This will include concepts such as anonymity, unlinkability, disclosure, transparency, and privacy by design.

1.1 Motivations for Data Privacy

We can distinguish three main motivations for companies to apply data privacy procedures. They are legislation, companies own interest, and avoiding privacy breaches. Let us look into these three motivations in more details.

Legislation

Nowadays privacy is a fundamental right that is protected at different levels. The Universal Declaration of Human Rights states in its Article 12 the following.

> **Article 12.** No one shall be subjected to arbitrary interference with his privacy, family, home or correspondence, nor to attacks upon his honour and reputation. Everyone has the right to the protection of the law against such interference or attacks. (Universal Declaration of Human Rights [11], 10 December 1948, UN General Assembly)

In Europe, the Council of Europe entered into force in 1953 the European Convention on Human Rights (ECHR). This convention contains the following article focusing on privacy.

Article 8. Right to respect for private and family life.

1. Everyone has the right to respect for his private and family life, his home and his correspondence.

2. There shall be no interference by a public authority with the exercise of this right except such as is in accordance with the law and is necessary in a democratic society in the interests of national security, public safety or the economic wellbeing of the country, for the prevention of disorder or crime, for the protection of health or morals, or for the protection of the rights and freedoms of others.

(European Convention on Human Rights [12], November 1950, Council of Europe)

State-members of the Council of Europe have the international obligation to comply with the convention (see e.g. [13], p. 14). There are 47 member states in January 2014.

In the European Union, in 25 May 2018 shall apply the General Data Protection Regulation (GDPR) that entered into force on 24 May 2016. The regulation consolidates some rights related to privacy as the right to erasure and the right to rectification.

In USA, the most relevant laws are probably the Health Insurance Portability and Accountability Act (HIPAA, 1996), the Patriot Act (2001), and the Homeland Security Act (2002).

Several states include similar rights in their constitutions, and laws have been approved to guarantee these rights. Information of national level legislation in Europe can be found in [13] and within the European Union in [14,15]. Reference [15] links to different studies and country reports describing and comparing several data protection laws (some non-European countries including USA, Japan, and Australia are also discussed). Reference [13] (Chap. 2) includes not only a review and analysis of legislation but also a dicussion on the data protection terminology.

Companies own interest

Companies and organizations need to define protocols and practices so that their activities with data are compliant with the laws. In addition to that, companies are often interested in protecting their own data because of their own interest (data releases can lead to the disclosure of information that can be used by competitors for their own profit). As we will see in Sect. 3.2, there is an area in data privacy (holder privacy) focused on this problem.

Avoiding privacy breaches

There are several cases of privacy breaches that have been reported in the media. Some of them are due to a bad understanding of data privacy fundamentals, and a lack of application of appropriate technologies. Others (a large majority) are due to security breaches (e.g. unencrypted data forgotten in public transport).

Two well known examples of voluntary data releases that lead to privacy issues are the AOL [16] and the Netflix [17] cases.

In the AOL case (August/September 2006), queries submitted to AOL by 650,000 users within a period of three months were released for research purposes. Only queries from AOL registered users were released. Numerical identifiers were published instead of real names and login information. Nevertheless, from the terms in the queries a set of queries was linked to people.

In the NETFLIX case, a database with the ratings to films of about 500,000 subscribers of Netflix was published in 2006 to foster research in recommender systems. About 100 milion film ratings were included in the database. Reference [17] used the Internet Movie Database as background knowledge to link some of the records in the database to known users.

There is a discussion on whether information leakages as in the two cases above should prevent the publication of information that is useful to people, and, in general, to what extent some risk should be tolerated.

There are some works stating that privacy is "impossible" [18,19] because either the disclosure risk is too high or the modifications we need to apply to the data to make them safe are too large (and make them useless). There are some people arguing against this position [20,21] and others stating that even if the risk exists, this should not be overemphasized.

For example, [22] states that "there is only a single known instance of de-anonymization for a purpose other than the demonstration of privacy". Then, she adds that "this is not surprising, because the marginal value of the information in a public dataset is usually too low to justify the effort for an intruder. The quantity of information available in the data commons is outpaced by the growth in information self-publicized on the Internet or collected for commercially available consumer data". In a later work, Jane Bambauer [23] compares the chance of breaking disclosure in the 2009 ONC Study (see [24]) which is 0.013% (2 over 15,000 HIPAA-compliant records) and the chance of dying from a motor vehicle accident that year in USA, which is 0.018% (18 over 100,000 licensed drivers in 2008 in USA see [25]). Thus, the probability of reidentification in the 2009 ONC Study is smaller than the one of dying from a motor vehicle accident. For the AOL case, the proportion of actual reidentified users is even smaller.

It is true that for high dimensional data (e.g. records with a large number of variables) it is in general difficult to make data safe without compromising their quality. This can also be the case of some small datasets. That is why privacy is a challenging area of research.

In any case, any release of sensitive data needs to evaluate the risk of disclosure. Data has to be correctly protected so that the risk is low, and this requires an accurate evaluation of the risk. A data set well protected may have a low risk, as the computations by Bambauer above show, but conversely, no protection (or a bad protection) may lead to a high risk. For comparison, Sweeney reports in [26] a set of experiments and show that "87.1% (216 million of 248 million) of the population in the United States had characteristics that were likely made them unique based only on *5-digit ZIP, gender, date of birth*", and that "3.7% of the population in the United States had characteristics that were likely made them unique based only on *5-digit ZIP, gender, Month and year of birth*".

1.2 Privacy and Society

It is well known that privacy is not only a technical problem but that has social roots as well. Schoeman edited a book [27] which reproduces a few articles devoted to social, legal, and philosophical questions related to privacy, and also wrote a book [28] that analyzes and summarizes some of the main issues on these topics. In the introduction of Schoeman's book [29], the author gives classifications and definitions of these topics. In particular, it discusses the nature of privacy (their definitions and whether it is a moral right even if there is no associated legal right) and whether it is culturally relative.

About the nature of privacy, Shoeman claims ([29], p. 2) that privacy has been defined according to different perspectives and distinguishes the following ones.

- Privacy as a claim, entitlement, or right of an individual to determine what information about himself or herself may be communicated to others.
- Privacy as the measure of control an individual has over: information about himself; intimacies of personal identity; or who has sensory access to him.
- Privacy as a state or condition of limited access to a person. A person has privacy to the extent that others have limited access to information about him.

Cultural relativeness makes us wonder whether the importance of privacy is the same among all people, if privacy is superfluous and dispensable as a social value. Another question is whether there are aspects of life which are inherently private or just conventionally so.

From a historical perspective, the first explicit discussion of privacy as a legal right is attributed to Warren and Brandeis in their 1890 paper [30]. Nevertheless, privacy issues were already considered before (see e.g. Stephen's paper in 1873 [31]). Warren and Brandeis argues that the right to privacy is the *next step which must be taken for the protection of the person, and for securing to the individual what Judge Cooley calls the right "to be let alone"*.

The causes of needing this right are, according to Warren and Brandeis, related to the inventions and business methods of that time (i.e., 1890).

Instantaneous photographs and newspaper enterprise have invaded the sacred precincts of private and domestic life; and numerous mechanical devices threaten to make good the prediction that "what is whispered in the closet shall be proclaimed from the house-tops." (...)
Gossip is no longer the resource of the idle and of the vicious, but has become a trade, which is pursued with industry as well as effrontery (...) To occupy the indolent, column upon column is filled with idle gossip, which can only be procured by intrusion upon the domestic circle.
(S.D. Warren and L.D. Brandeis, 1890 [30])

Computers, the internet, and online social networks have made the right to privacy even more relevant. The persistence and easy access to data caused the need for other related rights such as the right to rectification and the right to be forgotten. See e.g. the General Data Protection Regulation in the European Union.

The relationship between privacy, freedom and respect for people has been discussed in the literature among others by Benn in 1971 (see [32], also reproduced in [27]). See, for example:

> Anyone who wants to remain unobserved and unidentified, it might be said, should stay at home or go out only in disguise. Yet there is a difference between happening to be seen and having someone closely observe, and perhaps record, what one is doing, even in a public place. (...) Furthermore, what is resented is not being watched *tout court*, but being watched without leave. (S.I. Benn, 1971 [32], p. 225 in [27])

The dangers, and misuse, of this being watched without leave quickly relates the lack of privacy to Orwellian states [33] (and to the Panopticon). See e.g. the poem by the Galician poet Celso Emilio Ferreiro.

> Un grande telescopio nos vixía
> coma un ollo de Cíclope
> que sigue os nosos pasos
> e fita sin acougo o noso rumbo,
> dende tódalas fiestras,
> dende tódalas torres,
> dende tódalas voces que nos falan.
> (Celso Emilio Ferreiro, Longa noite de pedra, 1962 [34])

We finish this section with another quotation of S.I. Benn that expresses the relationship between privacy and autonomy.

> This last stage of my argument brings me back to the grounds for the general principle of privacy, to which I devoted the first half of this paper. I argued that respect for someone as a person, as a chooser, implied respect for him as one engaged on a kind of self-creative enterprise, which could be disrupted, distorted, or frustrated even by so limited an intrusion as watching. A man's view of what he does may be radically altered by having to see it, as it were, through another man's eyes. Now a man has attained a measure of success in his enterprise to the degree that he has achieved autonomy. To the same degree, the importance to him of protection from eavesdropping and Peeping Toms diminishes as he becomes less vulnerable to the judgements of others, more reliant on his own (though he will still need privacy for personal relations, and protection from the grosser kinds of persecution). (S.I. Benn, 1971 [32], p. 242 in [27])

1.3 Terminology

In this section we give a few definitions of concepts related to privacy. The main concepts we review are anonymity, linkability, disclosure, and pseudonyms. For these and related definitions, we refer the reader to [35], a document written by A. Pfitzmann and M. Hansen with the goal of fixing a common terminology in the

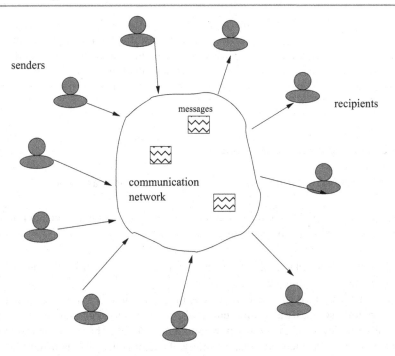

Fig. 1.1 Privacy setting in a communication network

area of privacy and privacy enhancement technologies, and to [36] which includes a glossary of terms in Statistical Disclosure Control. Both documents include a larger number of definitions than the ones included here. We focus on the ones that are directly related to the content of this book.

1.3.1 The Framework

Following Pfitzmann and Hansen [35] and most literature on information security, we consider a system in which there are senders and recipients which communicate through a communication network through messages (see Fig. 1.1). The senders are also called the actors, and the receivers the actees. There is no distinction on whether the senders (or the receivers) are human or not. That is, the same context applies when a user queries a database or a search engine.

The terms attacker, adversary, and intruder are used to refer to the set of entities working against some protection goal. Pfitzmann and Hansen [35] consider the terms attacker and adversary as synonyms, attacker being of older use and adversary is more recently used in information security research. Privacy preserving data mining uses the terms adversary and adversarial attacks (see e.g. [2]). Intruder is the most used term in statistical disclosure control.

The goal of adversaries are to increase their knowledge on the items of interest (e.g. subjects, messages, actions). Knowledge can be described in terms of probabilities on these items. So, we can represent adversaries increasing their knowledge on who sent a particular message, or on what is described in a record. Naturally, the more knowledge an intruder has, the more the probabilities are closer to the true probabilities.

1.3.2 Anonymity and Unlinkability

In this framework, anonymity and anonymity set are defined as follows.

Definition 1.1 [35] Anonymity of a subject means that the subject is not identifiable within a set of subjects, the anonymity set.

In this definition not identifiable means that the subject is not distinguishable from the other subjects within the anonymity set.

Note that here the quantification of the anonymity is implicit, as different sizes of the anonymity set may imply different levels of anonymity. For example, s_1 and s_2 have different levels of anonymity if the anonymity set of s_1 has only three subjects and the one of s_2 has ten subjects. Naturally, the anonymity of s_2 is better than the one of s_1.

Even in the case of two subjects with anonymity sets of the same cardinality, we may have different levels of identification. We illustrate this with the following example.

Example 1.1 Let $AS(s_1) = \{s_{11}, s_{12}, s_{13}\}$ be the anonymity set of s_1, and let us consider the following two probability distributions on this set: $P_1 = (1/3, 1/3, 1/3)$ and $P_2 = (0.9, 0.05, 0.05)$. They correspond to the probability that the correct identification of s_1 is s_{11}, s_{12}, or s_{13}. Then, with respect to anonymity, P_1 is better than P_2.

Because of that, we have different levels of identification according to the sizes of the anonymity set and to the shape of the probability distributions. A discussion on different probability distributions with respect to reidentification can be found in [37].

We revise below the previous definition of anonymity making explicit the level of identification.

Definition 1.2 From an adversary's perspective, anonymity of a subject s means that the adversary cannot achieve a certain level of identification for the subject s within the anonymity set.

Linkability is an important concept in privacy as its negation, unlinkability, implies anonymity. For example, when the interest of the attacker is to know which is the

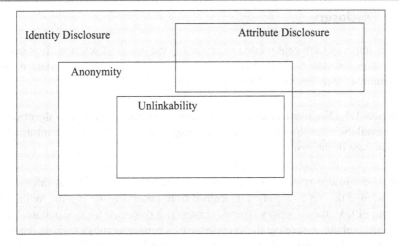

Fig. 1.2 Relationship between anonymity, unlinkability, and disclosure

sender or the recipient of a message, unlinkability with the sender implies anonymity of the sender. We define unlinkability below.

Definition 1.3 [35] Unlinkability of two or more items of interest (IOI) from an attacker's perspective means that within the system (comprising these and possibly other items), the attacker cannot sufficiently distinguish whether these IOIs are related or not.

Pfitzmann and Hansen [35] point out that unlinkability is a sufficient condition of anonymity but not a necessary condition. That is, unlinkability implies anonymity. We represent graphically in Fig. 1.2, the concepts introduced in this section as well as their relationships. The property that unlinkability implies anonymity is represented in Fig. 1.2 with the box of unlinkability inside the box of anonymity. On the contrary, we might have a case where linkability is possible but that anonymity is not compromised. As an example of the later case consider when an attacker can link all messages of a transaction, due to timing, nevertheless if all of them are encrypted and no information can be obtained about the subjects in the transactions, then anonymity is not compromised [35]. In Fig. 1.2 this example belongs to the region of the anonymity box outside the unlinkability box.

To finish this section, we include examples of anonymity in communications that are defined in terms of unlinkability, following [35]:

- **Sender anonymity**. It is not possible to link a certain message to the sender.
- **Recipient anonymity**. It is not possible to link a certain message to the receiver.
- **Relationship anonymity**. It is not possible to link a certain message to both the sender and the receiver.

All of them correspond to an appropriate definition of the set of items of interest.

1.3.3 Disclosure

Another important concept related to privacy is the one of disclosure. It is mainly used within the statistical disclosure control and privacy preserving data mining communities. Note that this concept does not appear in [35].

Definition 1.4 Disclosure takes place when attackers take advantage of the observation of available data to improve their knowledge on some confidential information about an item of interest.

Thus, disclosure is defined in terms of the additional confidential information or knowledge that an adversary can acquire from observing the system. Within the SDC and PPDM, the adversary typically observes a protected file or database. Then, disclosure is about improving the knowledge of a particular subject whose data is in the database. According to the terminology we introduce in Sect. 3.1.1, this subject is known as the respondent of the database.

According to [38,39], disclosure can be studied from two perspectives:

- **Identity disclosure**. This type of disclosure takes place when the adversary can correctly link a respondent to a particular record in the protected data set. Only respondents whose data have been published can be identified. According to Lambert in [38], even if "the intruder learns nothing sensitive from the identification, the re-identification itself may compromise the security of the data file". Identity disclosure is also known as entity disclosure. It corresponds to the concept of linkability discussed above.
- **Attribute disclosure**. This type of disclosure takes place when the adversary can learn something new about an attribute of a respondent, even when no relationship can be established between the individual and the data. That is, disclosure takes place when the published data set permits an intruder to increase his accuracy on an attribute of the respondent. This approach was first formulated in [5] (see also [40,41]).
  There are situations in which an intruder is able to determine whether the information about an individual is in a database or not. Just this information without learning anything else may cause a significant disclosure. For example, this is the case if the published database is the set of patients of a specialized hospital. This situation can also be seen as attribute disclosure.

It is important to note that identity disclosure is neither stronger nor weaker than attribute disclosure. As we show below, although some general rules apply, we can have any of them without the other.

The most common implication is that identity disclosure implies attribute disclosure. This is the case when we identify one of the records in the protected data set using a subset of the attributes. In this case, after identification, we learn about the values of the other attributes.

Table 1.1 Table that permits an adversary to achieve identity disclosure and attribute disclosure. Only variables City, Age, and Illness are published

Respondent	City	Age	Illness
ABD	Barcelona	30	Cancer
COL	Barcelona	30	Cancer
GHE	Tarragona	60	AIDS
CIO	Tarragona	60	AIDS
HYU	Tarragona	58	Heart attack

To illustrate this case, let us consider an adversary with the following information: $(HYU, Tarragona, 58)$. This indicates that the adversary has a friend HYU from Tarragona who is 58 years old. Then, let us assume that a hospital has published variables City, Age, and Illness of Table 1.1 of their patients (i.e., names in the first column are not published). Then, the adversary can link the data to the last row of the table and this results into identity disclosure. This disclosure implies attribute disclosure, as the adversary learns that HYU has had a heart attack.

However, in general we might have all possible cases.

- Neither identity nor attribute disclosure
- Identity disclosure and attribute disclosure
- No identity disclosure but attribute disclosure
- Identity disclosure but not attribute disclosure

We have discussed the second case. We illustrate below two other cases. Note that the last case is the one that makes the implication "identity disclosure implies attribute disclosure" fail.

- **Attribute disclosure without identity disclosure**. This is the case when we have sets of indistinguishable records, for example, due to k-anonymity (see Sects. 5.8 and 6.4).

  Note that when all indistinguishable records have the same value for a confidential attribute, we have attribute disclosure for this attribute. This issue motivated the definition of l-diversity, as we discuss in Sect. 5.8.

  To illustrate this case, let Table 1.2 be the data published by the hospital. As before, only attributes City, Age and Illness are published, and let the intruder have the following information $(ABD, Barcelona, 30)$. Using the published data the intruder will be able to infer that ABD has cancer while identity disclosure has not taken place. Note that the anonymity set for the information *(Barcelona, 30)* contains both ABD and COL.

- **Identity disclosure without attribute disclosure**. This is the case when all attributes are needed for reidentification, or when the attributes not used for reidentification do not cause disclosure.

Table 1.2 Table that permits an adversary to achieve attribute disclosure without identity disclosure. Only variables City, Age, and Illness are published

Respondent	City	Age	Illness
ABD	Barcelona	30	Cancer
COL	Barcelona	30	Cancer
GHE	Tarragona	60	AIDS
CIO	Tarragona	60	AIDS

Table 1.3 Table that permits an adversary to achieve identity disclosure without attribute disclosure. Only variables City, Age, and Illness are published

Respondent	City	Age	Illness
TTY	Manresa	60	AIDS
GTJ	Manresa	60	Heart attack
FER	Manresa	30	Heart attack
DRR	Barcelona	30	Heart attack
ABD	Barcelona	30	Cancer
COL	Barcelona	30	Cancer
GHE	Tarragona	60	AIDS
CIO	Tarragona	60	AIDS
HYU	Tarragona	60	Heart attack

To illustrate this case, let us consider the publication of attributes City, Age and Illness in Table 1.3 when the intruder has the following information (*HYU, Tarragona, 60, Heart attack*). In this case reidentification is possible, as the only possible record is the last one, but as all the attributes are used in the reidentification, no attribute disclosure takes place.

Note that identity disclosure and anonymity are exclusive. That is, identity disclosure implies non-anonymity, and anonymity implies no identity disclosure. Therefore, as in the case discussed above, we can have anonymity and attribute disclosure. In general, if all objects in the anonymity set satisfy a property p, the attacker can infer this property p for the subject of interest.

The relationship between all these terms is represented graphically in Fig. 1.2. Note that the region of identity disclosure (colored region) is the one outside the region of anonymity because they are exclusive.

We will discuss identity and attribute disclosure in more detail in Chap. 5. Nevertheless, it is important to point out that identity disclosure can be understood as a binary property. That is, an individual is reidentified or not. However, the case of attribute disclosure needs to be clarified, as it is not uncommon to consider that any release of data would lead to attribute disclosure to some extent. Otherwise, the utility of the protected data might be too low. See e.g. the discussion in [42] where the authors state:

> At the other extreme, any improvement in our knowledge about an individual could be considered an intrusion. The latter is particularly likely to cause a problem for data mining, as the goal is to improve our knowledge. (J. Vaidya et al., 2006 p. 7 [42])

Among the types of disclosure found in the literature, we also find inferential disclosure. We do not distinguish it from attribute disclosure, as in both cases the intruder gets some additional information on a particular attribute. The difference is whether the information is obtained directly from the protected data set, or whether it is inferred from this data set. The following example illustrates inferential disclosure. The example is taken from the term Inferential Disclosure in the Glossary of Statistical terms in [36].

Example 1.2 The data may show a high correlation between income and purchase price of a home. As the purchase price of a home is typically public information, a third party might use this information to infer the income of a data subject.

We will discuss the concept of disclosure in more details in Chap. 5.

An issue that [35] discusses repeatedly is the fact that anonymity cannot increase. The same happens with disclosure. Disclosure never decreases. Once some information has been released, this information can be used to compute an anonymity set and have some level of disclosure. Any additional information released later will eventually reduce the anonymity set and increase the disclosure.

1.3.4 Undetectability and Unobservability

In this section we discuss the concepts of undetectability and unobservability. We start with undetectability.

Definition 1.5 [35] Undetectability of an item of interest (IOI) from an attacker's perspective means that the attacker cannot sufficiently distinguish whether it exists or not.

For example, if attackers are interested in messages, undetectability holds when they cannot distinguish the messages in the system from random noise. Steganography [43, 44] gives tools to embed undetectable messages in other physical or digital objects for their transmission. In computer science, messages are often embedded in images but also in other objects such as databases [45].

Undetectability is not directly related to anonymity or privacy, but it is clear that an undetected message will not raise the interest of the attacker. Having said that, there are approaches to counter attack steganographic systems. This area of research is known as steganalysis. For example, [46] describes visual and statistical attacks for steganographic images.

Unobservability is a related term. It is defined as follows.

Definition 1.6 [35] Unobservability of an item of interest means

- undetectability of the IOI against all subjects uninvolved in it and
- anonymity of the subject(s) involved in the IOI even against the other subject(s) involved in that IOI.

So, unobservability pressumes undetectability but at the same time it also pressumes anonymity in case the items are detected by the subjects involved in the system. From this definition, it is clear that unobservability implies anonymity and undetectability.

1.3.5   Pseudonyms and Identity

Pseudonyms are related to anonymity and privacy. It is well known that their origin is previous to computers and internet, and in some cases they were already used to avoid the disclosure of sensitive information, such as gender. This is the case of Amandine Aurore Lucile Dupin (Paris, 1804—Nohant-Vic, 1876) who used the pen name George Sand, and of Caterina Albert (L'Escala, 1869—L'Escala, 1966) who used the pen name Víctor Català.

Similar uses are found nowadays in information systems and internet.

Definition 1.7 [35] A pseudonym is an identifier of a subject other than one of the subject's real names.

Definition 1.8 Pseudonymising is defined as the replacing of the name or other identifiers by a number in order to make the identification of the data subject impossible or substantially more difficult. (Federal Data Protection Act, Germany 2001; see p. 4 in [47])

The holder of a pseudonym corresponds to the subject to which the pseudonym refers to. Several pseudonyms may correspond to the same subject. See pseudonyms P and Q in Fig. 1.3. In addition, it is also possible that several subjects share the same pseudonym (a group pseudonym). In this latter case, the subjects may define an anonymity set. See pseudonym R in Fig. 1.3.

In some circumstances, pseudonyms permit to cover the range between anonymity (no linkability) to accountability (maximum linkability). An individual can have a set of pseudonyms to avoid linkability.

Nevertheless, in some systems it is easy to establish links between a pseudonym and its holder. See e.g. the case of user names in email accounts and social networks,

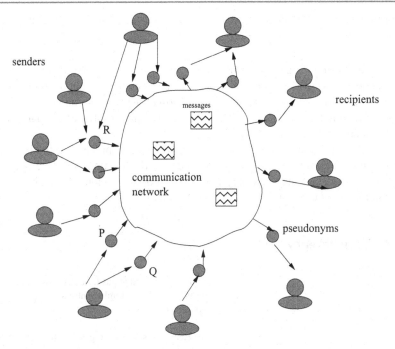

senders

recipients

messages

communication
network

pseudonyms

R

P

Q

Fig. 1.3 Pseudonyms in a communication network

access logs (as the case of AOL, see Sect. 1.1), and unique identifier number in software and services (e.g., the unique ID of the Chrome browser).

To complete this section we review the concept of identity, from the computer science perspective, and then some additional related concepts such as role and partial identity.

Definition 1.9 [35] An identity is any subset of attribute values of an individual person which sufficiently identifies this individual person within any set of persons. So usually there is no such thing as "the identity", but several of them.

Definition 1.10 Roles are defined as the set of actions that users (people) are allowed to perform.

For example, in a university, there are individuals who work as professors or members of the administration, and others who study. Therefore, there are people with the role of a professor, an official, or a student. In an information system for universities we would have implemented these three roles. Roles are not exclusive as we may have e.g. professors who are studying (hopefully not the subjects of their own professorship) and, because of that, information systems permit a user to login using different roles.

Definition 1.11 [35] Each partial identity represents the person in a specific context or role.

Then, the terms virtual (partial) identity and digital (partial) identity are used interchangeably to refer to the data stored by a computer-based application. Virtual identities can correspond to subjects' login information into a computer-based applications. Systems can then have profiles for each virtual (partial) identity. That is, the data for such identities.

Identity management focuses on creating, storing and processing the identities of users in an information system. This includes, authentication of a user, assigning roles to users so that they can only perform authorized actions, managing stored information about users and their logged activities.

1.3.6 Transparency

> Il faut qu'il n'exige pas le secret, et qu'il puisse sans inconvénient tomber entre les mains de l'ennemi
>
> A. Kerckhoffs, La cryptographie militaire, 1883 [48]

When data is published, it is a good practice to give details on how data has been produced. This naturally includes a description of any data protection process applied to the data as well as any parameter of the methods.

With respect to data privacy, we have maximum transparency when data is published with information on the protection procedure (masking method) applied together with its parameters (if any).

Transparency has a positive effect on data utility. If researchers know how data has been protected, they can take this information into account in data anaysis. As we will see in Sect. 6.1.3, in the case that masking is done by adding noise, information on the type of the added noise can be used to compute the exact values of the mean and variance of the original data.

Nevertheless, the information published can also be used by intruders to attack the data. In fact, it has been proven that effective attacks can be designed for some particular data protection methods (masking methods) if related information is also published (see e.g. [49–52]). We will discuss in Sects. 6.1.1 and 6.1.2 some attacks that take advantage of how data is protected. We call this type of attacks transparency attacks. The first explicit mention of the term transparency in data privacy was in [53], although the transparency attacks precede the use of this term.

A good protection method is one that is resilient to attacks even in the case of transparency. That is, a good data protection method is the one that cannot be attacked even in case an intruder has all the information about how data has been protected. Methods proposed in [49,54,55] were the first proposed with this objective.

This approach is similar to the Kerckhoffs's principle [48] in cryptography. The principle states that a cryptosystem should be secure even if everything about the system is public knowledge, except the key. We can formulate this principle for masking methods as follows.

Principle 1 *Given a privacy model, a masking method should be compliant with this privacy model even if everything about the method is public knowledge.*

Transparency attacks look for vulnerabilities of masking methods when the transparency principle is applied.

1.4 Privacy and Disclosure

From a computational perspective, privacy breaches are usually defined in terms of a type of disclosure. In this case, identity disclosure is the strongest disclosure.

Under the identity disclosure perspective, privacy problems appear when data about individuals make them unique. However, as we have already reported above (see e.g. [26]) only a few attributes (e.g., *zip code, gender and date of birth*) are enough to make individuals unique. In a more general context, individuals are multifaceted and each of their faces (or interests) do not lead to uniqueness (e.g., opera or rock listener, F.C. Barcelona or Skövde IK supporter, vegetarian or beef lover). It is the intersection of their interests that make people unique. See [56] and Sect. 4.2.2.2 for further discussion on this issue.

Building large enough anonymity sets permits us to ensure appropriate privacy levels with respect to identity disclosure. Data protection procedures correspond to different ways of achieving this.

The role of attribute disclosure in data privacy when identity disclosure does not take place is not so clear. This is so because attribute disclosure is often the fundamental reason for building statistical and data mining models. Nevertheless, some data protection procedures have been developed to avoid or reduce this type of disclosure.

1.5 Privacy by Design

Software needs to be developed so that privacy requirements are satisfied. "Privacy by design" was introduced by Ann Cavoukian to stress the fact that privacy "must ideally become an organization's default mode of operation" [57] and thus, not something to be considered a posteriori. In this way, privacy requirements need to be specified, and then software and systems need to be engineered from the beginning[1] taking these requirements into account.

Cavoukian established seven principles that may permit to fulfill the objectives of privacy by design. The principles are listed below.

1. Proactive not reactive; Preventative not remedial.
2. Privacy as the default setting.
3. Privacy embedded into design.
4. Full functionality—positive-sum, not zero-sum.
5. End-to-end security—full lifecycle protection.
6. Visibility and transparency—keep it open.
7. Respect for user privacy—keep it user-centric.

There is discussion on how privacy by design has to be put into practice, or, in other words, which are the design strategies that permit us to build systems that are compliant with our privacy requirements. Engineering strategies for implementing privacy by design have been discussed in e.g. [58–62].

Among these references, we underline [58,60]. Reference [60] states that a key strategy for achieving privacy is data minimization, and then establish some strategies based on minimization (minimize collection, minimize disclosure, minimize replication, minimize centralization, minimize linkability, and minimize retention). See the following quotation.

> After further examination of existing privacy preserving system designs, it became evident that a whole family of design principles are lumped under the term 'data minimization'. The term conceals a number of design strategies that experts apply intuitively when developing privacy preserving systems. A number of these are constraints on information flows like minimizing collection, disclosure, linkability, replication, retention and centrality. Systems engineered by applying these constraints intend to 'minimize risk' by avoiding a single point of failure, and minimize the need to trust data collectors and processors by putting data under the user's control. (S. Gürses et al., 2015 [60])

In contrast, [58] defines eight design strategies which are named as follows: minimise, separate, aggregate, hide, inform, control, enforce, and demonstrate. The author then discusses how these design strategies cover the privacy principles of the ISO 29100 Privacy framework: (i) consent and choice, (ii) purpose legitimacy and specification, (iii) collection limitation, (iv) data minimisation, (v) use, retention and disclosure limitation, (vi) accuracy and quality, (vii) openness, transparency and notice, (viii) individual participation and access, (ix) accountability, (x) information security, and (xi) privacy compliance.

---

[1] See e.g. "In the context of developing IT systems, this implies that privacy protection is a system requirement that must be treated like any other functional requirement. In particular, privacy protection (together with all other requirements) will determine the design and implementation of the system" in [58].

In this book we will not discuss (software) strategies and methodologies to satisfy these strategies and principles. We will focus on the technologies that permit us to build systems that satisfy the privacy by design requirements.

References

1. Uribe, K.: Bitartean heldu eskutik. Susa (2001)
2. Aggarwal, C.C., Yu, P.S. (eds.): Privacy-Preserving Data Mining: Models and Algorithms. Springer, New York (2008)
3. Willenborg, L., de Waal, T.: Elements of Statistical Disclosure Control. Lecture Notes in Statistics. Springer, New York (2001)
4. Domingo-Ferrer, J., Torra, V.: Disclosure control methods and information loss for microdata. In: Doyle, P., Lane, J.I., Theeuwes, J.J.M., Zayatz, L. (eds.) Confidentiality, Disclosure, and Data Access: Theory and Practical Applications for Statistical Agencies, North-Holland, pp. 91–110 (2001)
5. Dalenius, T.: Towards a methodology for statistical disclosure control. Statistisk Tidskrift **5**, 429–444 (1977)
6. Estivill-Castro, V., Brankovic, L.: Data swapping: balancing privacy against precision in mining for logic rules. In: Proceedings of the DaWaK 1999. LNCS, vol. 1676, pp. 389–398 (1999)
7. Agrawal, R., Srikant, R.: Privacy preserving data mining. In: Proceedings of the ACM SIGMOD Conference on Management of Data, pp. 439–450 (2000)
8. Samarati, P.: Protecting respondents' identities in microdata release. IEEE Trans. Knowl. Data Eng. **13**(6), 1010–1027 (2001)
9. Sweeney, L.: k-anonymity: a model for protecting privacy. Int. J. Unc. Fuzz. Knowl. Based Syst. **10**(5), 557–570 (2002)
10. Chaum, D.L.: Untraceable electronic mail, return addresses, and digital pseudonyms. Commun. ACM **24**(2), 84–88 (1981)
11. http://www.un.org/en/documents/udhr/. Accessed Jan 2017
12. http://www.echr.coe.int/Documents/Convention_ENG.pdf. Accessed Jan 2017
13. Handbook on European Data Protection Law, European Union Agency for Fundamental Rights, December 2013. http://echr.coe.int/Documents/Handbook_data_protection_ENG.pdf. Accessed Jan 2017
14. http://ec.europa.eu/justice/data-protection/index_en.htm. Accessed Jan 2017
15. http://ec.europa.eu/justice/data-protection/document/studies/. Accessed Jan 2017
16. Barbaro, M., Zeller, T., Hansell, S.: A face is exposed for AOL searcher no. 4417749. New York Times, 9 August 2006
17. Narayanan, A., Shmatikov, V.: Robust de-anonymization of large sparse datasets. In: Proceedings of the 2008 IEEE Symposium on Security and Privacy (SP 2008), pp. 111–125 (2008)
18. de Montjoye, Y.-A., Radaelli, L., Singh, V.K., Pentland, A.S.: Unique in the shopping mall: on the reidentifiability of credit card metadata. Science **347**, 536–539 (2015)
19. Jändel, M.: Anonymization of personal data is impossible in practice, presented in Kistamässan om Samhällssäkerhet 2015 (2015)
20. Sánchez, D., Martínez, S., Domingo-Ferrer, J. (2016) Comment on "Unique in the shopping mall: reidentifiability of credit card metadata". Science 18 March 1274-a
21. Barth-Jones, D., El Emam, K., Bambauer, J., Cavoukioan, A., Malin, B.: Assessing data intrusion threats. Science **348**, 194–195 (2015)
22. Yakowitz, J.: Tragedy of the data commons. Harward J. Law Technol. **25**(1), 1–67 (2011)

23. Bambauer, J.: Tragedy of the deidentified data commons: an appeal for transparency and access. In: Joint UNECE/Eurostat Work Session on Statistical Data Confidentiality, Ottawa, Canada, 28–30 Oct 2013 (2013)
24. Lafky, D.: The Safe Harbor Method of De-Identification: An Empirical Test, Dept of Health and Human Servs. Office of the Nat'l Coordinator for Health Information Technical, 15–19 (2009). http://www.ehcca.com/presentations/HIPAAWest4/lafky_2.pdf
25. https://www.census.gov/compendia/statab/2012/tables/12s1103.pdf
26. Sweeney, L.: Simple Demographics Often Identify People Uniquely, Carnegie Mellon University, Data Privacy Working Paper 3. Pittsburgh 2000 (1997)
27. Schoeman, F.D.: Philosophical Dimensions of Privacy: An Anthology. Cambridge University Press, Cambridge (1984)
28. Schoeman, F.D.: Privacy and Social Freedom. Cambridge University Press, Cambridge (1992)
29. Schoeman, F.D.: Privacy: philosophical dimensions of the literature, Reproduced as Chap. 1 in [27] (adapted from Privacy: philosophical dimensions, American Philosophical Quaterly 21, 1984) (1984)
30. Warren, S.D., Brandeis, L.: The Right to Privacy, Harvard Law Review IV:5, 15 Dec 1890. http://groups.csail.mit.edu/mac/classes/6.805/articles/privacy/Privacy_brand_warr2.html
31. Stephen, J.F.: Liberty, Equality and Fraternity. Henry Hold and Co, London (1873)
32. Benn, S.I.: Privacy, freedom, and respect for persons. Reproduced as Chap. 8 in [27] (1971)
33. Orwell, G.: Nineteen Eighty-four, A Novel. Harcourt, Brace, New York (1949)
34. Ferreiro, C.E.: Longa noite de pedra. Xerais (1990)
35. Pfitzmann, A., Hansen, M.: A terminology for talking about privacy by data minimization: anonymity, unlinkability, undetectability, unobservability, pseudonymity, and identity management (2010). http://dud.inf.tu-dresden.de/literatur/Anon_Terminology_v0.34.pdf
36. Glossary of Statistical terms, Inferential disclosure. http://stats.oecd.org/glossary/detail.asp?ID=6932 (version 2005), http://neon.vb.cbs.nl/casc/glossary.htm (version 2009). Accessed Jan 2017. Also published as a glossary in [213]
37. Stokes, K., Torra, V.: n-Confusion: a generalization of k-anonymity. In: Proceedings of the Fifth International Workshop on Privacy and Anonymity on Information Society (PAIS 2012) (2012)
38. Lambert, D.: Measures of disclosure risk and harm. J. Official Stat. **9**, 313–331 (1993)
39. Paass, G.: Disclosure risk and disclosure avoidance for microdata. J. Bus. Econ. Stat. **6**, 487–500 (1985)
40. Duncan, G.T., Lambert, D.: Disclosure-limited data dissemination. J. Am. Stat. Assoc. **81**, 10–18 (1986)
41. Duncan, G.T., Lambert, D.: The risk disclosure for microdata. J. Bus. Econ. Stat. **7**, 207–217 (1989)
42. Vaidya, J., Clifton, C.W., Zhu, Y.M.: Privacy Preserving Data Mining. Springer, New York (2006)
43. Wayner, P.: Disappearing Cryptography, Information Hiding: Steganography & Watermarking, 3rd edn. Morgan Kaufmann Publishers, Amsterdam (2009)
44. Min, W., Liu, B.: Multimedia Data Hiding. Springer, New York (2003)
45. Bras-Amorós, M., Domingo-Ferrer, J.: On overlappings of digitized straight lines and shared steganographic file systems. Trans. Data Priv. **1**(3), 131–139 (2008)
46. Westfeld, A., Pfitzman, A.: Attacks on steganographic systems: breaking the steganographic utilities EzStego, Jsteg, Steganos, and S-Tools-and some lessons learned In: LNCS, vol. 1768, pp. 61–76 (2000)
47. Korff, D.: Country report on different approaches to new privacy challenges in particular in the light of technological developments. Country studies.In: A.4—Germany, European Commission. Directorate-General Justice, Freedom and Security (2010). http://ec.europa.eu/justice/data-protection/document/studies/index_en.htm
48. Kerckhoffs, A.: La cryptographie militaire. J. Sci. Militaires **IX**, 5–38 (1883)

49. Nin, J., Herranz, J., Torra, V.: Rethinking rank swapping to decrease disclosure risk. Data Knowl. Eng. **64**(1), 346–364 (2007)
50. Nin, J., Herranz, J., Torra, V.: On the disclosure risk of multivariate microaggregation. Data Knowl. Eng. **67**(3), 399–412 (2008)
51. Nin, J., Torra, V.: Analysis of the univariate microaggregation disclosure risk. New Gener. Comput. **27**, 177–194 (2009)
52. Winkler, W.E.: Single ranking micro-aggregation and re-identification, Statistical Research Division report RR 2002/08 (2002)
53. Karr, A.F.: The role of transparency in statistical disclosure limitation. In: Joint UNECE/Eurostat Work Session on Statistical Data Confidentiality (2009)
54. Domingo-Ferrer, J., Torra, V.: Towards fuzzy c-means based microaggregation. In: Grzegorzewski, P., Hryniewicz, O., Gil, M.A. (eds.), Soft Methods in Probability and Statistics, pp. 289–294 (2002)
55. Domingo-Ferrer, J., Torra, V.: Fuzzy microaggregation for microdata protection. J. Adv. Comput. Intell. Intell. Inf. **7**(2), 153–159 (2003)
56. Juárez, M., Torra, V.: Toward a privacy agent for information retrieval. Int. J. Intel. Syst. **28**(6), 606–622 (2013)
57. Cavoukian, A.: Privacy by design. The 7 foundational principles in Privacy by Design. Strong privacy protection—now, and well into the future (2011). https://www.ipc. on.ca/wp-content/uploads/Resources/7foundationalprinciples.pdf, https://www.ipc.on.ca/wp-content/uploads/Resources/PbDReport.pdf
58. Hoepman, J.-H.: Privacy design strategies. In: Proceedings of the IFIP SEC 2014, pp. 446–459 (2014)
59. Gürses, S., Troncoso, C., Diaz, C.: Engineering privacy by design. Comput. Priv. Data Prot. **14**, 3 (2011)
60. Gurses, S., Troncoso, C., Diaz, C.: Engineering privacy by design reloaded. In: Proceedings of the Amsterdam Privacy Conference 2015 (2015)
61. Danezis, G., Domingo-Ferrer, J., Hansen, M., Hoepman, J.-H., Le Métayer, D., Tirtea, R., Schiffner, S.: Privacy and data protection by design—from policy to engineering, ENISA Report (2014)
62. D'Acquisto, G., Domingo-Ferrer, J., Kikiras, P., Torra, V., de Montjoye, Y.-A., Bourka, A.: Privacy by design in big data: an overview of privacy enhancing technologies in the era of big data analytics, ENISA Report (2015)

Machine and Statistical Learning

2

> It may be summed up in one short sentence:
> 'Advise your son not to purchase your wine in Cambridge'
>
> C. Babbage, Passages from the life of a philosopher, 1864, p. 25 [1]

In this chapter we will review some analytical tools. We will need them to analyse the effect of data protection methods on the data. We do not distinguish here whether the tools described belong to the statistics community or to the machine learning comunity.

Having said that, there is some discussion in the literature on the similarities and differences between machine learning and statistics (see e.g. [2–7]). As a personal comment, and following [4,5], I consider that there is a large overlapping that is increasing year by year due to the research directions in artificial intelligence and machine learning. But of course, some particular topics do not belong to the overlapping region. This is the case of e.g. inductive logic programming and official statistics. The first topic falls within the area of machine learning and the second in the area of statistics.

Some argue that machine learning focuses on prediction without focusing too much on the underlying distribution of the data. For example, [7] illustrates this with the case of data with more variables than samples. This type of problem has been studied in machine learning for e.g. recommender systems, and it is difficult to build parametric models for this type of data. Observe however the study of predictive inference [8] within statistics.

Others [5] mention as a main difference the size of the data, being machine learning devoted to large data sets. From my point of view, this is not an issue of machine

V. Torra, *Data Privacy: Foundations, New Developments and the Big Data Challenge*, Studies in Big Data 28,
DOI 10.1007/978-3-319-57358-8_2

learning, but of data mining, where issues such as dimensionality reduction and scalability are of extreme importance. Data mining has its origin on commercial data where files and databases are typically of large dimensions (although not as large as today's big data). In addition, current research on statistics (see e.g. [9–11]) also consider high-dimensional data and large data sets.

In any case, in real applications, techniques originating from both, machine learning and the statistics communities are needed. As this chapter is mainly instrumental, we describe the tools without distinguishing their origin or their development. In order to present the different tools, we will mainly follow the terminology of Hastie, Tibshirani, and Friedman in [12].

2.1 Classification of Techniques

Methods and techniques in machine learning are typically classified into three large classes according to the type of information available. They are supervised learning, unsupervised learning, and reinforcement learning. We discuss them below. All of them presume a set of labeled examples X used in the learning process. We will also presume that each example x_i in X is described in terms of a set of attributes $A_1, \ldots, A_M$. We will use $A_j(x_i)$ to denote the value of the attribute A_j for example x_i. That is, x_i is a vector in an M-dimensional space.

- **Supervised learning**. In this case, it is presumed that for each example in X there is a distinguished attribute A_y. The goal of supervised learning algorithms is to build a model of this attribute with respect to the other attributes. For example, if the attributes $A_1 \ldots A_M$ are numerical and the distinguished attribute A_y is also numerical, the goal of the supervised learning algorithm might be to express A_y as a linear regression of $A_1, \ldots, A_k$ for $k \leq M$. In general, A_y is expressed as a function of $A_1, \ldots, A_k$. When A_y is categorical, we call this attribute the class. Figure 2.1 summarizes the notation for supervised learning.

  For the sake of simplicity, it is usual to use x_i to denote $A(x_i)$, X to denote the full matrix (or just the full matrix without the attribute A_y), and Y to denote the column A_y. In some applications we need some training sets C that are subsets of X. That is, $C \subseteq X$. Then, we denote by M_C that we have a model learnt from C.

  Formally, let us consider a training set C defined in terms of the examples X where for each example x_i in X, we have its known label (outcome or class label) y. Then, we presume that we can express y in terms of a function f of x_i and some

Fig. 2.1 Notation for supervised machine learning

	A_1	$\ldots$	A_M	A_y
x_1	$A_1(x_1)$	$\ldots$	$A_M(x_1)$	$y_1 = A_y(x_1)$
$\vdots$	$\vdots$		$\vdots$	$\vdots$
x_N	$A_1(x_N)$	$\ldots$	$A_M(x_N)$	$y_N = A_y(x_N)$

error. That is, $y = f(x_i) + \varepsilon$. With this notation we can say that the goal is to build a model M_C that depends on the training set C such that $M_C(x_i)$ approximates $f(x_i)$ for all x_i in the training set. This model is then used to classify unseen instances.

Within supervised learning algorithms we investigate regression problems and classification problems. They are described further below.

- **Regression problems**. They correspond to the case in which A_y is numerical. Models include linear regression (i.e., models of the form $A_y = \sum_{i=1}^{k} a_i A_i + a_0$ for real numbers a_i), non-linear regression, and neural networks.
- **Classification problems**. They correspond to the case in which A_y is categorical. Models include logistic regression, decision trees, and different types of rule based systems.
- **Search problems**. They correspond to the problems in artificial intelligence to speed up search algorithms. However, this type of problems are of no interest in this book.

- **Unsupervised learning**. In this case, all the attributes are equal with respect to the learning process, and there is no such a distinguishable attribute. In this case, algorithms try to discover patterns or relationships in the data that can be of interest. Unsupervised learning includes clustering and association rules mining. The former discovers partitions in the data and the latter discovers rules between the attributes.
- **Reinforcement learning**. In this case we presume that there is already a system (or model) that approximates the data. When the model is used, the system receives a reward if the outcome of the model is successful and a penalty if the outcome is incorrect. These rewards are used to update the model and increase its performance.

2.2 Supervised Learning

A large number of algorithms for supervised machine learning have been developed. We give a brief overview of a few of them. For details and further algorithms the reader is referred to [12, 13].

2.2.1 Classification

We explain superficially decision trees and the nearest neighbor.

2.2.1.1   Decision Trees

A decision tree classifies an element x by means of a chain of (usually binary) questions. These questions are organized as a tree with the first question in the top (the root) and classes in the leaves.

Machine learning algorithms build decision trees from data. The goal is to classify new elements correctly, and minimize the height of the tree (i.e., minimize the number of questions to be asked when a new element arrives).

2.2.1.2   Nearest Neighbor

Classification of a new example x is based on finding the nearest record from a set of stored records (the training set C). The class of this record is returned. Formally,

$$class(x) = class(arg \min_{x' \in C} d(x', x))$$

where $class(x) = A_y(x)$ using the notation given above.

An alternative is to consider the k nearest records and then return the class of the majority of these k records. This approach corresponds to the k-nearest neighbor.

2.2.2   Regression

There are different approaches to build models for regression. In this section we only review the expressions for linear regression models. There are, however, alternatives. For example, we have non-linear regression models and we can use k-nearest neighbor for regression. The k-nearest neighbor for regression follows the approach of the k-nearest neighbor explained above but instead of returning the class of the majority, the mean of the output of the majority is used.

Let us now focus on linear regression. We will give the expressions in matrix form. For details on regression see e.g. [14].

We denote the data (the training set) by the pair X, Y where Y corresponds to the variable to be modeled (the dependent variable) and X corresponds to the variables of the model (the independent or explanatory variables). In linear regression the model has the following form

$$y_i = \beta_0 + \beta_1 x_{i1} + \beta_2 x_{i2} + \cdots + \beta_M x_{iM} + \varepsilon_i.$$

In matrix form, using

$$Y^T = (y_1 y_2 \ldots y_N)$$

$$\beta^T = (\beta_0 \beta_1 \beta_2 \ldots \beta_M)$$

and

$$X = \begin{pmatrix} 1 & x_{11} & x_{12} & \dots & x_{1M} \\ 1 & x_{21} & x_{22} & \dots & x_{2M} \\ \vdots & \vdots & \vdots & & \vdots \\ 1 & x_{N1} & x_{N2} & \dots & x_{NM} \end{pmatrix} \tag{2.1}$$

we have that the model has this form:

$$Y = X\beta + \varepsilon.$$

Then, the ordinary least squares (OLS) method estimates the parameters β of the model computing

$$\beta = [X^T X]^{-1} X^T Y.$$

Statistical properties of this method (as e.g. the Gauss-Markov theorem) can be found in [14].

2.2.3 Validation of Results: k-Fold Cross-Validation

This is one of the most used approaches to evaluate the performance of a model. The approach is based on having a data set Z and then building several pairs of (*training, testing*) sets from this single data set. For each pair we can compute a model using the training set and evaluate its performance with the test set.

For a given parameter k, we divide the set Z into k subsets of equal size. Let

$$Z = (Z_1, Z_2, \dots, Z_k)$$

be these sets. Then, we define for $i = 1, \dots, k$ the pair of training and test sets (C_i^{Tr}, C_i^{Ts}) as follows:

$$C_i^{Tr} = \cup_{j \neq i} Z_j$$

$$C_i^{Ts} = Z_i.$$

Given these sets, we can compute the accuracy of any machine learning algorithm that when applied to the training set C returns a classification model M_C using the following expression:

$$accuracy = \frac{\sum_{i=1}^{k} |\{x | M_c(x) = A_y(x), x \in C_i^{Ts}\}|}{\sum_{i=1}^{k} |C_i^{Ts}|}.$$

Note that accuracy is not the only way to evaluate the performance of a classifier. Nevertheless, we will not discuss alternatives here. Cross-validation can also be used to evaluate regression.

2.3   Unsupervised Learning

The area of unsupervised learning has developed several families of methods to extract information from unclassified raw data. In this section we will focus on methods for clustering, for association rule mining, and on the expectation-maximization algorithm. They are the ones that will be used later in this book.

2.3.1   Clustering

> The objective is not to choose a 'best' clustering technique or program. Such a task would be fruitless and contrary to the very nature of clustering.
>
> Dubes and Jain, 1976 [15], p. 247

The goal of clustering, also known as cluster analysis, is to detect the similarities between the data in a set of examples. Different cluster methods differ on the type of data considered and on the way used to express the similarities.

For example, most clustering methods are applicable to numerical data. Nevertheless, other methods can be used on categorical data, time series, search logs, or even on nodes in social networks. With respect to the way used to express the similarities between the data, some clustering methods build partitions of the data objects, others build fuzzy partitions of these data, fuzzy relationships between the objects, and hierarchical structures (dendrograms).

In all cases, the goal of a clustering method is to put similar objects together in the same group or cluster, and put dissimilar ones in different clusters. For achieving this, a crucial point is how to measure the similarity between objects. Different definitions of similarity and distance lead to different clusters.

Methods and algorithms for clustering can be classified according to several dimensions. As expressed above, one is the type of data being clustered, another is the type of structure built around the data. Reference [12] (page 507) considers another dimension that refers to our assumptions on data. The following classes of clustering algorithms are considered: combinatorial algorithms, algorithms for mixture modeling, and algorithms that are mode seekers. See the outline in Fig. 2.2. We briefly describe these classes below.

- **Combinatorial algorithms**. They do not presume any underlying probability distribution. They directly work on the data.
- **Mixture modeling algorithms**. They presume an underlying probability density function. Assuming a parametric approach, clustering consists of finding the parameters of the model (a mixture of density functions). E.g., two Gaussian distributions are fitted to a set of points.

Clustering methods

- Combinatorial methods.

  - Partitive methods (top-down): c-means, fuzzy c-means.
  - Agglomerative methods (bottom-up): single linkage.

- Mixture modeling methods.
- Mode seeker methods.

Fig. 2.2 A classification of some clustering methods

- **Mode seeker algorithms**. They also presume an underlying probability density function but in this case the perspective is nonparametric. So, there is no such a prior assumption that data follows a particular model.

In the rest of this section we review some methods for clustering. We focus on methods for numerical data that lead to crisp and fuzzy partitions. These algorithms belong to the family of combinatorial algorithms. Both type of methods are *partitive*, this means that we have initially a single set of data (a single cluster) and then we partition this cluster into a set of other clusters. In contrast, we find in the literature agglomerative methods that start with as many clusters as data, and then merge some of these clusters to build new ones. Agglomerative methods can be seen as bottom-up methods, and partitive methods as top-down.

Following Dubes and Jain [15,16], we can distinguish between clustering methods (or techniques) and clustering algorithms (or programs). A clustering method is to specify the general strategy for defining the clusters. In contrast, a clustering algorithm implements the strategy and might use some heuristics. This difference will be further stressed below when describing the k-means.

2.3.1.1 Crisp Clustering

Given a data set X a crisp clustering algorithm builds a partition of the objects in X. Formally, $\Pi = \{\pi_1, \ldots, \pi_c\}$ is a partition of X if $\cup \pi_i = X$ and for all $i \neq j$ we have $\pi_i \cap \pi_j = \emptyset$.

For any set of n objects, given c, the number of possible partitions of c clusters is the Stirling number of the second kind (see [16] p. 91, and [12] Sect. 14.30):

$$S(n, c) = \frac{1}{c!} \sum_{k=1}^{c} (-1)^{c-k} \binom{c}{k} k^n.$$

When c is not known and any number of clusters of $c = 1, \ldots, n$ is possible, the number of possible partitions of a set with n elements is the Bell number.

$$B_n = \sum_{k=1}^{n} S(n, c).$$

It is known (see [17]) that for $n \in \mathbb{N}^n$

$$\left(\frac{n}{e \ln n}\right)^n < B_n < \left(\frac{0.792n}{\ln(n+1)}\right)^n.$$

Different methods exist for selecting or constructing one of these partitions. In optimal clustering, the partition is selected as the one that minimizes an objective function. That is, given an objective function OF, and a space of solutions S, select Π as the solution s that minimizes OF. Formally,

$$\Pi = \arg\min_{s \in S} OF(s).$$

One of the most used methods for clustering is k-means, also known as crisp c-means in the community of fuzzy clustering. This algorithm uses as inputs the data set X and also the number of clusters c. This method is defined as an optimal clustering with the following objective function.

$$OF(\Pi) = \sum_{k=1}^{c} \sum_{x \in \pi_k} ||A(x) - p_k||^2 \tag{2.2}$$

Here, π_k, which is a part of partition Π, corresponds to a cluster and p_k is the centroid or prototype of this cluster. $||u||$ is the norm of the vector u. That is, $||u|| = \sqrt{u_1^2 + \ldots u_M^2}$.

Expression 2.2 can be rewritten in terms of characteristic functions χ_k of sets π_k. That is, for each set π_k we have a characteristic function $\chi_k : X \rightarrow \{0, 1\}$ such that $\chi_k(x) = 1$ if and only if $x \in \pi_k$. Using this notation, the goal of the clustering algorithm is to determine the set of characteristic functions $\chi = \{\chi_1, \ldots, \chi_c\}$ as well as the cluster centroids $P = \{p_1, \ldots, p_c\}$.

The characteristic functions define a partition. Because of that we require χ to satisfy

- $\chi_k(x) \in \{0, 1\}$ for all $k = 1, \ldots, c$ and $x \in X$, and that
- for all $x \in X$ there is exactly one k_0 such that $\chi_{k_0}(x) = 1$.

The last condition can be equivalently expressed as $\sum_{k=1}^{c} \chi_k(x) = 1$ for all $x \in X$.
Taking all this into account, we formalize the c-means problem as follows:

Minimize

$$OF(\chi, P) = \sum_{k=1}^{c} \sum_{x \in X} \chi_k(x)||A(x) - p_k||^2 \tag{2.3}$$

subject to

$$\chi \in M_c = \left\{\chi_k(x) | \chi_k(x) \in \{0, 1\}, \sum_{k=1}^{c} \chi_k(x) = 1 \text{ for all } x \in X\right\}$$

This optimization problem is usually solved by means of an iterative algorithm that interleaves two steps. In the first step, we presume that P is known and determines the partition χ that minimizes the objective function $OF(\chi, P)$ given P. In the second step, we presume that the partition χ is known and we determine the cluster centers P that minimize the objective function $OF(\chi, P)$ given χ. This process is repeated until convergence.

This algorithm does not ensure a global minimum, but ensures convergence to a local minimum. We discuss this in more detail later.

Let us now formalize the steps above and give expressions for their calculation. The steps are as follows.

Step 1.   Define an initial partition and compute its set of centroids P.
Step 2.   Solve $min_{\chi \in M_c} OF(\chi, P)$.
Step 3.   Solve $min_P OF(\chi, P)$.
Step 4.   Repeat steps 2 and 3 till convergence.

The solution of Step 2 consists of assigning each object in X to the nearest cluster. Formally, for all $x \in X$ use the following assignments.

- $k_0 := \arg\min_i ||A(x) - p_i||^2$
- $\chi_{k_0}(x) := 1$
- $\chi_j(x) := 0$ for all $j \neq k_0$

Note that in this definition k_0 depends on $x \in X$.

To prove that this is the optimal solution of the problem stated in Step 2, let us consider the objective function

$$OF(\chi, P) = \sum_{k=1}^{c} \sum_{x \in X} \chi_k(x) ||A(x) - p_k||^2.$$

Naturally, we have that for a given x and $p_1, \ldots, p_c$, it holds

$$||A(x) - p_k|| \geq ||A(x) - p_{k_0}||$$

for all $k \in \{1, \ldots, c\}$, when k_0 is the index $k_0 = \arg\min_i ||A(x) - p_i||^2$. Therefore, the assignment $\chi_{k_0}(x) = 1$ and $\chi_j(x) = 0$ for all $j \neq k_0$ minimizes

$$\sum_{x \in X} \chi_k(x) ||A(x) - p_k||^2$$

for all $k \in \{1, \ldots, c\}$, and thus the objective function.

The solution of Step 3 consists in computing for all $k = 1, \ldots, c$.

$$p_k = \frac{\sum_{x \in X} \chi_k(x) A(x)}{\sum_{x \in X} \chi_k(x_i)} \tag{2.4}$$

To prove that this is the optimal centroid we consider again the objective function $OF(\chi, P)$ and derive it with respect to p_k. Taking into account that $\frac{\partial OF}{\partial p_k} = 0$, we obtain an expression for p_k. Note that

$$0 = \frac{\partial OF}{\partial p_k} = 2 \sum_{x \in X} \chi_k(x)(A(x) - p_k)(-1),$$

and, therefore, we get the equation

$$-2 \sum_{x \in X} \chi_k(x) A(x) + \sum_{x \in X} \chi_k(x) p_k = 0,$$

that leads to Eq. 2.4.

If we put all the items together, we get Algorithm 1.

As stated above, there is no guarantee that this algorithm leads to the global optimal solution. However, it can be proven that it converges to a local optimal one. Note that at each step the objective function is reduced. In Step 2, with fixed centroids P, the objective function is reduced changing χ. Then, in Step 3, with fixed χ, the objective function is reduced changing P. As the objective function is always positive, convergence is ensured.

Different executions of this algorithm using the same initialization lead to the same results. Nevertheless, due to the fact that the algorithm does not ensure a global

---

Algorithm 1: Clustering: c-means.

Step 1.   Define an initial partition and compute its centroid P.

Step 2.   Solve $min_{\chi \in M_c} OF(\chi, P)$ as follows:

- For all $x \in X$,

  - $k_0 := \arg\min_i \|A(x) - p_i\|^2$
  - $\chi_{k_0}(x) := 1$
  - $\chi_j(x) := 0$ for all $j \in \{1, \ldots, c\}$ s.t. $j \neq k_0$

Step 3.   Solve $min_P OF(\chi, P)$ as follows:

- for all $k \in \{1, \ldots, c\}$,

  - $p_k := \frac{\sum_{x \in X} \chi_k(x) A(x)}{\sum_{x \in X} \chi_k(x_i)}$

Step 4.   Repeat steps 2 and 3 till convergence

---

minimum but a local one, we have the situation where different initializations can lead to different local minima. This fact is very important when we need to compare clusters obtained from the application of this algorithm.

To partially solve this problem, we can use some of the existing methods for selecting a good initialization for a dataset X. For some initialization methods, see e.g. [18]. Another option is to apply the same algorithm several times to the same data set X, but with different initializations. Then, each application will lead to a partition with its corresponding value for the objective function. Let $r = 1, \dots, R$ denote the rth application, Π_r the partition obtained and OF_r its corresponding objective function. All partitions Π_r are local optima of the same objective function OF. Then, we select the partition Π_r with minimum OF_r. That is, we select the partition

$$r_0 = \arg\min OF_r.$$

This approach does not ensure finding the global optimum, it can still lead to a local optimum. Nevertheless it gives us more chances of finding it. We have used this approach in [19], where 20 different executions were used, and in [20] where 30 were used.

The outcome of c-means permits us to define classification rules for any element d in the same domain D of the elements in X. That is, not only the elements x can be classified but any $d \in D$ can be classified to one of the clusters $\pi_1, \dots, \pi_c$. The classification rule is:

$$cluster(d) = \arg\min_{k=1}^{c} ||d - p_k||^2.$$

The application of this classification rule in a domain D results into a Voronoi diagram described by the centers P and the Euclidean distance. Recall that the Voronoi diagram of a domain D divides D into a set of regions. Here, the regions are $(R_k)_{k \in \{1,\dots,c\}}$, where

$$R_k = \{d \in D| \; ||d - p_k|| \leq ||d - p_j|| \text{ for all } j \neq k\}.$$

2.3.1.2 Fuzzy Clustering

Fuzzy clustering algorithms return a fuzzy partition instead of a crisp partition. In fuzzy partitions, clusters typically overlap. This causes elements $x \in X$ to have partial membership to different clusters. Partial membership is represented by a value in the [0, 1] interval.

In this section we review some of the algorithms that lead to fuzzy partitions. We begin by reviewing the notion of membership function used to define fuzzy sets [21], and then the notion of fuzzy partition. For a discussion on the difference between fuzzy and probabilistic uncertainty (from a *fuzzy* point of view) see [22].

Definition 2.1 [21]
Let X be a reference set. Then $\mu : X \rightarrow [0, 1]$ is a membership function.

Definition 2.2 [23] Let X be a reference set. Then, a set of membership functions $\mathcal{M} = \{\mu_1, \ldots, \mu_c\}$ is a fuzzy partition of X if for all $x \in X$ we have

$$\sum_{i=1}^{c} \mu_i(x) = 1$$

Fuzzy c-means (FCM) [24] is one of the most used algorithms for fuzzy clustering. It can be seen as a generalization of crisp c-means that has a similar objective function. The solution of the problem is a fuzzy partition. That is, given a value c, the algorithm returns c membership functions $\mu_1, \ldots, \mu_c$ that define a fuzzy partition of the elements of the domain X. Figure 2.3 discusses a naive fuzzification of c-means.

The notation follows the one of c-means. X is the set of records, $P = \{p_1, \ldots, p_c\}$ representing the cluster centers or centroids, μ_i is the membership function of the ith cluster and, then, $\mu_i(x_k)$ is the membership of the kth record to the ith cluster. μ_{ik} is also used as an expression equivalent to $\mu_i(x_k)$.

Fuzzy c-means has two parameters. One is the number of clusters c, as in the c-means. Another is a value m that measures the degree of fuzziness of the solution. The value m should be larger than or equal to one. When $m = 1$, the problem to optimize corresponds to the c-means and the algorithm returns a crisp partition. Then, the larger the m, the fuzzier is the solution. In particular, for large values of m, we have that the solutions are completely fuzzy and memberships in all clusters are $\mu_i(x_k) = 1/c$ for all i and $x_k \in X$.

The optimization problem follows.

Minimize
$$OF_{FCM}(\mu, P) = \{\sum_{i=1}^{c} \sum_{x \in X} (\mu_i(x))^m \|x - p_i\|^2\}$$
subject to   (2.5)
$$\mu_i(x) \in [0, 1] \text{ for all } i \in \{1, \ldots, c\} \text{ and } x \in X$$
$$\sum_{i=1}^{c} \mu_i(x) = 1 \text{ for all } x \in X.$$

This problem is usually solved using Algorithm 2. The algorithm is an iterative process similar to the one of c-means. It iterates two steps. One step estimates the membership functions of elements to clusters (taking centroids as fixed). The other step estimates the centroids for each cluster (taking membership functions as fixed). The algorithm converges but as in the case of c-means the solution can be a local

---

Naive fuzzy c-means. A naive fuzzification of the c-means algorithm is to replace the constraint of χ in $\{0, 1\}$ in Equation 2.3 by another requiring χ to be a value in $[0, 1]$. Nevertheless, this fuzzification has no practical effect. It does not lead to fuzzy solutions. In other words, all solutions of this alternative problem are crisp partitions. That is, although χ is permitted to take values different to 0 and 1, all solutions have values of χ in the extremes of the interval $[0,1]$.

Fig. 2.3 Remark on a naive fuzzy c-means

Algorithm 2: Clustering: fuzzy c-means.

Step 1.   Generate initial P

Step 2.   Solve $min_{\mu \in M} O F_{FCM}(\mu, P)$ by computing for all $i \in \{1, \ldots, c\}$ and $x \in X$:

$$\mu_i(x) := \left(\sum_{j=1}^{c} \left(\frac{||x - p_i||^2}{||x - p_j||^2} \right)^{\frac{1}{m-1}} \right)^{-1}$$

Step 3.   Solve $min_P O F_{FCM}(\mu, P)$ by computing for all $i \in \{1, \ldots, c\}$:

$$p_i := \frac{\sum_{x \in X} (\mu_i(x))^m x}{\sum_{x \in X} (\mu_i(x))^m}$$

Step 4.   If the solution does not converge, go to Step 2; otherwise, stop.

---

optimum. The algorithm does not discuss the case of denominators equal to zero. This is solved with adhoc definitions for μ (see e.g. [25–27]).

Expressions for $\mu_i(x)$ and p_i in Steps 2 and 3 are determined using Lagrange multipliers (see e.g. [24]). The expression to minimize includes the objective function $O F_{FCM}$ as well as the constraints $\sum_{i=1}^{c} \mu_i(x) = 1$ for all $x \in X$. Each constraint is multiplied by the corresponding Lagrange multiplier λ_k (for $k = 1, \ldots, N$).

$$L = O F_{FCM}(\mu, P) + \sum_{k=1}^{N} \lambda_k \left(\sum_{i=1}^{c} \mu_i(x_k) - 1 \right)$$
$$= \sum_{i=1}^{c} \sum_{k=1}^{N} (\mu_i(x))^m ||x - p_i||^2 + \sum_{k=1}^{N} \lambda_k \left(\sum_{i=1}^{c} \mu_i(x_k) - 1 \right) \tag{2.6}$$

Now, in order to find the expression for $\mu_i(x_k)$, we consider the partial derivatives of L with respect to $\mu_i(x_k)$, that need to be zero. These partial derivatives are

$$\frac{\partial L}{\partial \mu_i(x_k)} = m(\mu_i(x_k))^{m-1} ||A(x_k) - p_i||^2 + \lambda_k = 0$$

Therefore, we have the following expression for $\mu_i(x_k)$

$$\mu_i(x_k) = \left(\frac{-\lambda_k}{m ||A(x_k) - p_i||^2} \right)^{\frac{1}{m-1}}$$

Now, taking advantage of the fact that $\sum_{i=1}^{c} \mu_i(x) = 1$ for all $x \in X$, we get rid of λ_k and obtain the expression for $\mu_i(x_k)$ in Step 2.

Similarly, in order to find the expression for p_i, we proceed with the partial derivative of L with respect to p_i. That is,

$$\frac{\partial L}{\partial p_i} = \sum_{k=1}^{N} (\mu_i(x_k))^m 2 (A(x_k) - p_i) (-1) = 0.$$

From this expression we get the expression for p_i in Step 3.

2.3.1.3 Variations for Fuzzy Clustering

There are several variations of fuzzy c-means. One of them is entropy based fuzzy c-means (EFCM). This method, which was proposed in [28], introduces fuzziness into the solution by adding to the objective function a term based on entropy. In a way similar to fuzzy c-means the algorithm uses a parameter λ ($\lambda \geq 0$) to control the degree of fuzziness. The larger the parameter, the more crisp is the solution. When $\lambda \rightarrow \infty$, the added term becomes negligible and the algorithm corresponds to standard c-means. In contrast, the near the parameter λ is to zero, the fuzzier is the solution. For λ near to zero, solutions have memberships $\mu_i = 1/c$ for all $i = 1, \ldots, c$.

The optimization problem for EFCM is defined as follows.

Minimize
$$OF_{EFCM}(\mu, P) = \sum_{x \in X} \sum_{i=1}^{c} \{\mu_i(x)||x - p_i||^2 + \lambda^{-1}\mu_i(x)log\mu_i(x)\}$$
s.t.

$$\mu_i(x) \in [0, 1]$$
$$\sum_{i=1}^{c} \mu_i(x) = 1 \text{ for all } x \in X.$$

(2.7)

As in the previous algorithms, this problem is solved by an iterative process that repeats two steps. One finds membership values ($\mu_i(x)$) that minimize the objective function given centers, and the other that find centers (p_i) given membership values. The expressions follows.

$$p_i = \frac{\sum_{x \in X} \mu_i(x)x}{\sum_{x \in X} \mu_i(x)}$$

(2.8)

$$\mu_i(x) = \frac{e^{-\lambda||x - p_i||^2}}{\sum_{j=1}^{c} e^{-\lambda||x - p_j||^2}}$$

(2.9)

The last expression for $\mu_i(x)$ can be rewritten as follows.

$$\mu_i(x) = \frac{1}{1 + \frac{\sum_{j \neq i}^{c} e^{-\lambda||x - p_j||^2}}{e^{-\lambda||x - p_i||^2}}}$$

(2.10)

There are several variations of the algorithms described above. One of them was introduced in [29]. Fuzzy c-means has an implicit assumption that all clusters have equal size. Because of that, given two clusters, with cluster centers p_1 and p_2 the mid-point between the two centers has equal membership to both clusters. That is, $\mu_1((p_1 + p_2)/2) = \mu_2((p_1 + p_2)/2) = 0.5$. If one of the clusters is larger than the other, we might have some elements incorrectly classified. This is illustrated in Fig. 2.4. There is one cluster with 1000 points centered in $(-2, 0)$ and another cluster with 10 points centered in $(2, 0)$. Classification according to fuzzy c-means assigns all points (x, y) with $x \leq 0$ to the cluster centered in $(-2, 0)$ and all points with $x > 0$ to the cluster centered in $(2, 0)$. This will incorrectly assign some of the

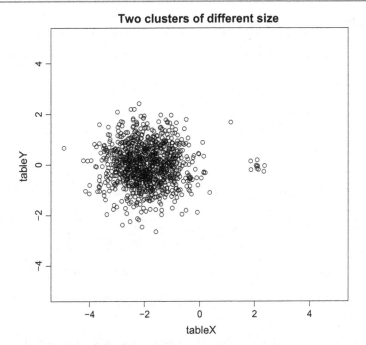

Fig. 2.4 Two clusters of different size. Fuzzy c-means and entropy based fuzzy c-means assigns some of the points to the incorrect cluster

points of the cluster in the left to the one in the right. The data represented in the figure was generated according to independent Normal distributions $N(0, 0.8)$ on each variable. One cluster with mean $(-2, 0)$ and the other with mean $(2, 0)$.

The algorithm presented in [29] solves this problem introducing a variable for each cluster that corresponds to its size. Then, the algorithm determines the variables corresponding to the sizes as well as the membership functions and the cluster centers. A similar approach was also introduced in [30].

Let us consider the same notation above, and let α_i denote the size of the ith cluster for $i \in \{1, \ldots, c\}$. Then, the variable size fuzzy c-means problem corresponds to the following optimization problem.

Minimize

$$O\,F_{VSFCM}(\alpha, \mu, P) = \sum_{i=1}^{c} \alpha_i \sum_{x \in X} (\alpha_i^{-1} \mu_i(x))^m ||x - p_i||^2$$

subject to

$$\mu_i(x) \in [0, 1] \text{ for all } i \in \{1, \ldots, c\} \text{ and } x \in X$$
$$\sum_{i=1}^{c} \mu_i(x) = 1 \text{ for all } x \in X$$
$$\sum_{i=1}^{c} \alpha_i = 1$$
$$\alpha_i \geq 0 \text{ for all } i = 1, \ldots, c.$$

$$(2.11)$$

Note that now the size of the ith cluster weights the contribution of each cluster in the objective function. The constraints are the same we had before, adding two constraints for the new variables α.

Algorithm 3: Clustering: variable size fuzzy c-means.

Step 1.   Generate initial α and P

Step 2.   Solve $min_{\mu \in M} OF_{VSFCM}(\alpha, \mu, P)$ by computing for all $i = 1, \ldots, c$ and $x \in X$:

$$\mu_i(x) := \Big(\sum_{j=1}^{c}\Big(\frac{\alpha_j}{\alpha_i}\Big)\Big(\frac{||x - p_i||^2}{||x - p_j||^2}\Big)^{\frac{1}{m-1}}\Big)^{-1}$$

Step 3.   Solve $min_P OF_{VSFCM}(\alpha, \mu, P)$ by computing for all $i = 1, \ldots, c$:

$$p_i := \frac{\sum_{x \in X}(\mu_i(x))^m x}{\sum_{x \in X}(\mu_i(x))^m}$$

Step 4.   Solve $min_\alpha OF_{VSFCM}(\alpha, \mu, P)$ by computing for all $i = 1, \ldots, c$:

$$\alpha_i := \Big(\sum_{j=1}^{c}\Big(\frac{\sum_{x \in X}(\mu_j(x))^m||x - p_j||^2}{\sum_{x \in X}(\mu_i(x))^m||x - p_i||^2}\Big)^m\Big)^{-1}$$

Step 5.   If the solution does not converge, go to Step 2; otherwise, stop.

This optimization problem is also solved using an iterative procedure that, in this case, iterates three steps. We have two that, as the ones we had in fuzzy c-means, compute membership values and cluster centers, and we have an additional step about the computation of the values α_i. The expressions for μ, p and α are given in Algorithm 3. Note that the expressions for p does not change, and that now the membership values depend on the values α.

The problem discussed above with the size of the clusters also appears in the case of entropy based fuzzy c-means. Because of that, a variable size entropy based fuzzy c-means was also defined in [28]. The optimization problem considers the following objective function.

$$OF_{EFCM}(\alpha, \mu, P) = \sum_{x \in X}\sum_{i=1}^{c}\mu_i(x)||x - p_i||^2 + \lambda^{-1}\mu_i(x)log(\alpha_i^{-1}\mu_i(x))$$

with the same constraints we had in variable size fuzzy c-means. In this case the iterative solution leads to the following expressions. Note that the expression for the centroids has not changed.

$$\mu_i(x) = \frac{\alpha_i e^{-\lambda||x - p_i||^2}}{\sum_{j=1}^{c}\alpha_j e^{-\lambda||x - p_j||^2}}$$

$$\alpha_i = \frac{\sum_{x \in X}\mu_i(x)}{|X|}$$

$$p_i = \frac{\sum_{x \in X} \mu_i(x) x}{\sum_{x \in X} \mu_i(x)}$$

2.3.1.4 Fuzzy Clustering and Noisy Data

In this section we mention a few more fuzzy clustering algorithms that have as their main characteristics that attempt to deal with the problem of noise.

Dave introduced in 1991, see [31], the Noise clustering method to reduce the effects of noisy data in the clusters obtained by FCM. The algorithm defines a noise cluster in order to assign noisy data to it. The distance of any element to this cluster is constant. This constant is a parameter of the algorithm (parameter δ). The other parameters are the ones of the fuzzy c-means. That is, the number of clusters c and the fuzziness m.

Later, Krishnapuram and Keller [32] introduced the Possibilistic c-means (PCM). While in fuzzy c-means, entropy fuzzy c-means and in all their variations the memberships of one element to clusters add up to one, this is not a requirement in the possibilistic c-means. In this case, values are only required to be in the [0, 1] interval. This fact implies that the distribution of memberships defines a possibility distribution, and, thus, a possibilistic interpretation is possible.

As [32] points out, a possibilistic solution cannot be achieved restating the optimization problem of the fuzzy c-means (Eq. 2.5) by just removing the constraint $\sum_{i=1}^{c} \mu_i(x) = 1$. Note that in this case the minimum is obtained with $\mu_i(x) = 0$ for all x and i. To avoid this problem, the authors remove the constraint and add an extra term to $OF_{FCM}(\mu, P)$ that depends on a parameter for each class: v_i for $i = 1, \ldots, c$. This parameter, using authors' words, "determines the distance at which the membership value of a point in a cluster becomes 0.5". So, in some sense this is related to the size of the cluster and similar to the variables α_i in variable size cluster. Note that while α_i are determined by the algorithm, v_i is a parameter of the algorithm. Nevertheless, the authors suggest an expression to compute v_i from data that has some resemblances with the expression for computing α_i in Algorithm 3.

The fact that v_i determines the distance where a membership of 0.5 is achieved implies that this value can be considered as a threshold value and that data with a distance to the centroid larger than v_i is considered as noise for that cluster.

In applications, this algorithm tends to locate the centroids in dense regions, and this causes that clusters have a large overlapping. This problem has been reported in [33,34]. The fuzzy possibilistic c-means (FPCM) is a variation of this algorithm proposed in [35] to avoid coincident clusters and to make the final clusters less sensitive to initializations. This algorithm was further improved in [36] where the authors of FPCM and PCM introduced possibilistic-fuzzy c-means (PFCM).

We review below the possibilistic-fuzzy c-means (PFCM). In the fuzzy c-means we have that the solutions are such that memberships to clusters add up to one; in the possibilistic fuzzy c-means this is not the case but memberships are only required to be positive. According to [32,36] this second type of memberships can be understood as the typicality of the element to the cluster. Then, possibilistic-fuzzy c-means includes both values membership and typicality. We will use μ to represent the usual membership and T to represent the set of typicalities, and $t_i(x)$ to represent the typicality of x to cluster i. The solution (the degree of possibility) for an element x will be $a(\mu_i(x))^m + b(t_i(x))^\eta$ with μ and t representing the degree of fuzziness of membership and typicality, respectively.

Values a and b are given and correspond, respectively, to the importance of membership and typicality. Naturally, when $a = 1$ and $b = 0$, we would have only membership as in the case of FCM.

The definition is not restricted to the Euclidean distance as previously. Now, the definition is in terms of an inner product norm, and thus other distances defined on inner product norms are valid. We will use $||X||_A = \sqrt{x^t A x}$ to denote any inner product norm based on matrix A.

Taking all this into account, the possibilistic-fuzzy c-means is defined by the following optimization problem.

Minimize
$$OF_{PFCM}(\mu, T, P) = \sum_{x \in X} \sum_{i=1}^{c} \left(a(\mu_i(x))^m + b(t_i(x))^\eta \right) \times ||x - p_i||_A^2$$
$$+ \sum_{i=1}^{c} \gamma_i \sum_{x \in X} (1 - t_i(x))^\eta$$
subject to   (2.12)
$$\sum \mu_i(x) = 1 \text{ for all } x \in X$$
$$0 \le \mu_i(x) \text{ for all } x \in X \text{ and all } i = \{1, \ldots, c\}$$
$$t_i(x) \le 1 \text{ for all } x \in X \text{ and all } i = \{1, \ldots, c\}$$

The parameters of the clustering method are a, b, m, η and γ_i for $i = \{1, \ldots, c\}$. In this algorithm, a and b define, as described above, the relative importance of fuzzy membership and typicality values in the objective function. We require $a > 0$ and $b > 0$. Then, m is the degree of fuzziness in the solution corresponding to the fuzzy c-means (we require $m > 1$), and η is the fuzzifier corresponding to the PCM (we also require $\eta > 1$). Then, the parameters $\gamma_i > 0$ for $i = 1, \ldots, c$ correspond to the parameters v_i of PCM (i.e., "the distance at which the membership value of a point in a cluster becomes 0.5" [32]).

The original paper gives some hints about the definition of the parameters γ_i. One of them is to bootstrap the algorithm using the fuzzy c-means and then define γ_i as follows for a $K > 0$ (with $K = 1$ the most common choice).

$$\gamma_i = K \frac{\sum_{x \in X} (\mu_i(x))^m D_{ikA}^2}{\sum_{x \in X} (\mu_i(x))^m}$$

where $D_{ikA} = ||x_k - v_i||_A$.

It is clear that when $b = 0$, and also $\gamma_i = 0$ for all i, the problem reduces to a FCM. Reference [36] also proves that also when $b = 0$, the problem implicitly becomes equivalent to FCM. With $a = 0$ it corresponds to PCM.

This optimization problem is solved using an iterative algorithm that interleaves the following three equations.

Let $D_{ikA} = ||x_k - v_i||_A$, then

$$u_{ik} = \left(\sum_{j=1}^{c} \left(\frac{D_{ikA}}{D_{jkA}} \right)^{\frac{2}{(m-1)}} \right)^{-1}$$

for all $1 \leq i \leq c$ and $1 \leq k \leq n$,

$$t_{ik} = \left(\sum_{j=1}^{n} \left(\frac{D_{ikA}}{D_{ijA}} \right)^{\frac{2}{(\eta-1)}} \right)^{-1}$$

for all $1 \leq i \leq c$ and $1 \leq k \leq n$, and

$$v_i = \frac{\sum_{k=1}^{n} (u_{ik}^m + t_{ik}^\eta) x_k}{\sum_{k=1}^{n} (u_{ik}^m + t_{ik}^\eta)}$$

for all $i = 1, \ldots, c$.

The algorithm can be bootstrapped using a set of initial cluster centers. The algorithm requires first to fix the parameters a, b, m, η, γ_i.

2.3.1.5 Comparison of Cluster Results

The need to compare different sets of clusters on the same data set appears naturally when we want to analyse the result of clustering algorithms. We can either need to compare the results of the clustering algorithm with respect to a known set of clusters (a reference partition), or to compare different executions of the same clustering algorithm.

Rand [37] (1971, Section 3), Hubert and Arabie [38], and Anderson et al. [39] mention the following applications related to clustering where functions to compare clusters are needed.

- **Comparison with a reference partition or golden standard**. This golden standard are the "natural" clusters using Rand's terminology. The results of a clustering algorithm are compared with the reference partition.
- **Comparison with noisy data**. The results of clustering the original data and the perturbed data permits us to measure the sensitivity of an algorithm to noisy data.
- **Comparison with supressed data (to missing individuals)**. Comparison of an original data set X and the same data set after suppression measures the sensitivity of an algorithm to missing data.
- **Comparison of two algorithms**. The results of two different algorithms applied to the same data are compared.
- **Comparison of two successive partitions given by the same algorithm**. This is useful for defining stopping criteria in iterative algorithms.

Table 2.1 Definitions for comparing two partitions Π and Π'.

	$I_\Pi(x_1) = I_\Pi(x_2)$	$I_\Pi(x_1) \neq I_\Pi(x_2)$	Total
$I_{\Pi'}(x_1) = I_{\Pi'}(x_2)$	r	t	$r + t = np(\Pi')$
$I_{\Pi'}(x_1) \neq I_{\Pi'}(x_2)$	s	u	$s + u$
	$r + s = np(\Pi)$	$t + u$	$\binom{|X|}{2}$

- **Comparison for prediction.** This is about using one of the partitions to predict the other.

In the remaining part of this section we discuss some of the distances and similarity measures for clusters that can be found in the literature. In our definitions we use Π and Π' to denote two partitions of a data set X. We presume that both partitions have the same number of parts, and that this number is n. Then, let $\Pi = \{\pi_1, \ldots, \pi_n\}$ and $\Pi' = \{\pi'_1, \ldots, \pi'_n\}$, that is, π_i and π'_i denotes a part of Π and, therefore, $\pi_i \subseteq X$ and $\pi'_i \subseteq X$ for all $i = 1, \ldots, n$.

Let $I_\Pi(x)$ denote the cluster of x in the partition Π. Then, let us define r, s, t, and u as follows:

- r is the number of pairs (x_1, x_2), with x_1 and x_2 elements of X, and both in the same cluster in Π and also both in the same cluster in Π'. That is, r is the cardinality of the set

$$\{(x_1, x_2) \text{ with } x_1 \in X, x_2 \in X, \text{ and } x_1 \neq x_2 | I_\Pi(x_1) = I_\Pi(x_2) \text{ and } I_{\Pi'}(x_1) = I_{\Pi'}(x_2)\};$$

- s is the number of pairs (x_1, x_2) where x_1 and x_2 are in the same cluster in Π but not in Π'. That is,

$$\{(x_1, x_2) \text{ with } x_1 \in X, x_2 \in X, \text{ and } x_1 \neq x_2 | I_\Pi(x_1) = I_\Pi(x_2) \text{ and } I_{\Pi'}(x_1) \neq I_{\Pi'}(x_2)\};$$

- t is the number of pairs where x_1 and x_2 are in the same cluster in Π' but not in Π. That is,

$$\{(x_1, x_2) \text{ with } x_1 \in X, x_2 \in X, \text{ and } x_1 \neq x_2 | I_\Pi(x_1) \neq I_\Pi(x_2) \text{ and } I_{\Pi'}(x_1) = I_{\Pi'}(x_2)\};$$

- u is the number of pairs where x_1 and x_2 are in different clusters in both partitions.

In addition, we denote $np(\Pi)$ as the number of pairs within clusters in the partition Π. That is

$$np(\Pi) = |\{(x_1, x_2)|x_1 \in X, x_2 \in X, x_1 \neq x_2, I_{\Pi(x_1)} = I_{\Pi(x_2)}\}|$$

where $|\cdot|$ is the cardinality of the set.

Note that using the notation above, $np(\Pi) = r + s$ and $np(\Pi') = r + t$. Table 2.1 presents a summary of these definitions.

The literature presents a large number of indices and distances to compare (crisp) partitions. The most well known are the Rand, the Jaccard and the Adjusted Rand index. We present these three and a few other ones. See e.g. [39, 40] for other indices.

- **Rand Index**. Defined by Rand in [37], this index is defined as follows:

$$RI(\Pi, \Pi') = (r + u)/(r + s + t + u)$$

For any Π and Π', we have $RI(\Pi, \Pi') \in [0, 1]$, with $RI(\Pi, \Pi) = 1$.

- **Jaccard Index**. It is defined as follows:

$$JI(\Pi, \Pi') = r/(r + s + t)$$

For any Π and Π', we have $JI(\Pi, \Pi') \in [0, 1]$, with $RI(\Pi, \Pi) = 1$.

- **Adjusted Rand Index**. This is a correction of the Rand index so that the expectation of the index for partitions with equal number of objects is 0. This adjustment was done assuming generalized hypergeometric distribution as the model of randomness. That is, if we consider a random generation of two partitions so that they have both n sets, the adjusted Rand index is zero. The definition of the index is as follows.

$$ARI(\Pi, \Pi') = \frac{r - exp}{max - exp}$$

where

$$exp = (np(\Pi)np(\Pi'))/(n(n-1)/2)$$

and where

$$max = 0.5(np(\Pi) + np(\Pi')).$$

First discussion of the adjusted Rand index is due to Morey and Agresti [41] and current expression is due to Hubert and Arabie [38].

- **Wallace Index**. This index is defined in terms of the following expression.

$$WI(\Pi, \Pi') = r/\sqrt{np(\Pi)np(\Pi')}$$

- **Mántaras Distance**. This distance proposed in [42] is defined for two partitions by

$$MD(\Pi, \Pi') = \frac{I(\Pi/\Pi') + I(\Pi'/\Pi)}{I(\Pi' \cap \Pi)}$$

where

$$I(\Pi/\Pi') = -\sum_{i=1}^{n} P(\pi_i') \sum_{j=1}^{n} P(\pi_j/\pi_i') \log P(\pi_j/\pi_i')$$

$$I(\Pi' \cap \Pi) = -\sum_{i=1}^{n} \sum_{j=1}^{n} P(\pi_i' \cap \pi_j) \log P(\pi_i' \cap \pi_j)$$

and where $P(A)$ is the probability of A estimated as $P(A) = |A|/|X|$.

Transaction number	Itemsets purchased	Items (only first letter)
x_1	{apple, biscuits, chocolate, doughnut, ensaïmada, flour}	{a, b, c, d, e, f}
x_2	{apple, biscuits, chocolate}	{a, b, c}
x_3	{chocolate, doughnut, ensaïmada}	{c, d, e}
x_4	{biscuits}	{b}
x_5	{chocolate, doughnut, ensaïmada, flour}	{c, d, e, f}
x_6	{biscuits, chocolate, doughnout}	{b, c, d}
x_7	{ensaïmada}	{e}
x_8	{chocolate, flour}	{c, f}

Fig. 2.5 Database $\mathscr{D}$ with 8 transactions for the itemset $I = \{$apple, biscuits, chocolate, doughnut, ensaïmada, flour, grapes$\}$

In the case of fuzzy clusters, a comparison can be done through α-cuts. That is, all those elements with a membership larger than α have their membership assigned to the value 1, others assigned to zero. Nevertheless, this process does not generate in general partitions and the above expressions cannot be used. Instead, we can compute the absolute distance between memberships. For binary memberships, this distance corresponds to the Hamming distance.

Alternatively, there are a few definitions to generalize some of the existing distances for crisp partitions to the case of fuzzy partitions. First attempts can be found in [43,44] and more consolidated work can be found in [39,45].

2.3.2 Association Rules

Association rules establish relationships between attributes or items in a database. Association rule learning algorithms try to find relevant rules in a given database. A typical application of these algorithms is market basket analysis. A market basket is the set of items a costumer buys in a single purchase. Then, a rule relates a set of some items that can be purchased with some other items that consumers usually buy at the same time.

Formally, a database $matchscr D = \{x_1, \ldots, x_N\}$ has N transactions (or records) consisting each one in a subset of a predefined set of items. Let $I = \{I_1, \ldots, I_m\}$ be the set of items. Then, $x_i \subset I$ for all $i \in 1, \ldots, N$. We call itemset any subset of I. Thus, x_i are itemsets.

In order to simplify the algorithms, it is usual to presume that the items in I are ordered in a particular preestablished order, and that the itemsets are not empty (i.e., $|x_i| \geq 1$).

An association rule is an implication of the form

$$X \Rightarrow Y$$

where X and Y are nonempty itemsets with no common items. Formally, $X, Y \subseteq I$ such that $|X| \geq 1$, $|Y| \geq 1$, and $X \cap Y = \emptyset$. X is called the antecedent and Y the consequent of the rule.

We review now some definitions that are needed later. We give examples for each definition based on the database $\mathcal{D}$ in Fig. 2.5.

- **Matching.** An itemset S *matches* a transaction T if $S \subseteq T$. For example, $S_1 = \{chocolate, doughnut\}$ matches transactions x_1, x_3, x_5, x_6, $S_2 = \{flour\}$ matches transactions x_1, x_5, and x_8, and there is no transaction matching $S_3 = \{grapes\}$.
- **Support count.** The support count of an itemset S, expressed by *Count(S)*, is the number of transactions (or records) that match S in the database $\mathcal{D}$. For example, $Count_{\mathcal{D}}(S_1) = 4$, $Count_{\mathcal{D}}(S_2) = 3$, and $Count_{\mathcal{D}}(S_3) = 0$.
- **Support.** The support of an itemset S, expressed by *Support(S)* is the proportion of transactions that contain all items in S in the database $\mathcal{D}$. That is, $Support_{\mathcal{D}}(S) = Count_{\mathcal{D}}(S)/|\mathcal{D}|$. For example,

$$Support_{\mathcal{D}}(S_1) = 4/8, \; Support_{\mathcal{D}}(S_2) = 3/8, \; \text{and} \; Support_{\mathcal{D}}(S_2) = 0/8.$$

When the context makes clear the database used, we will simply use $Count(s)$ and $Support(s)$ instead of $Count_{\mathcal{D}}(s)$ and $Support_{\mathcal{D}}(s)$.

Lemma 2.1 *Let I_1 and I_2 be two itemsets; then if $I_1 \subseteq I_2$ then*

$$Support(I_1) \geq Support(I_2).$$

Note that this holds because I_1 matches more itemsets in the database (because has less requirements) than I_2.

Given a rule of the form $R = (X \Rightarrow Y)$, its support will be the proportion of transactions in which both X and Y hold. This is computed defining the support of the rule as the support of the union of the two itemsets that define the rule.

- **Support of a rule.** The support of the rule $R = (X \Rightarrow Y)$ is the support of $X \cup Y$. I.e.,

$$Support(R) = Support(X \cup Y).$$

For example, the support of the rule

$$R_0 = (X \Rightarrow Y)$$

with $X = \{chocolate, doughnut\}$ and $Y = \{ensaïmada\}$ is

$$Support(X \Rightarrow Y) = Support(X \cup Y) = Support(\{chocolate,$$
$$doughnut, ensaïmada\}) = 3$$

because $\{chocolate, doughnut, ensaïmada\}$ matches x_1, x_3, and x_5.

Note that the rule R_0 in the last example does not hold for all the transactions in the database. Although the support of $X = \{chocolate, doughnut\}$ includes x_1, x_3, x_5, x_6, we have that x_6 does not include ensaïmada. Therefore, the rule does not hold for this transaction. In general, rules do not hold 100% of the time.

In order for a rule to be interesting, it should

- apply to a large proportion of records in the database, and
- have a large prediction capability.

For measuring the first aspect we can use the support. Note that support precisely measures the proportion of itemsets where the rule is applicable and holds. For the example above R_0 applies to

$$Support(R_0) = 3/8$$

of the transactions in $\mathcal{D}$.

For measuring the second aspect, we need to estimate the predictive accuracy of a rule. This is measured by the *confidence* of a rule, which is defined in terms of the support of the rule with respect to the support of the antecedent of the rule. In other words, the confidence states how many times the rule leads to a correct conclusion.

- **Confidence of a rule**. The confidence of the rule $R = (X \Rightarrow Y)$ is the support of the rule divided by the support of X. That is,

$$Confidence(R) = Support(X \cup Y)/Support(X).$$

Or, equivalently, using *Count*:

$$Confidence(R) = Count(X \cup Y)/Count(X).$$

As stated above, it is usual that rules are not exact. That is, $Confidence(R) < 1$ because for Lemma 2.1, $Support(X \cup Y) \leq Support(X)$.

As an example, the confidence of the rule $R_0 = (X \Rightarrow Y)$ is

$$\begin{aligned} Confidence(X \Rightarrow Y) &= Support(X \cup Y)/Support(X) \\ &= Support(\{c, d, e\})/Support(\{c, d\}) = 3/4. \end{aligned} \tag{2.13}$$

In order to filter the rules that are not interesting, we will reject all the rules that have a support below a certain threshold. That is, we reject all the rules that only apply to a small set of transactions. For example, the rules with a support less than 0.01. Given a threshold for the support $thr - s$, we say that an itemset I_0 is supported when $Support(I_0) \geq thr - s$.

Algorithm 4: Association Rule Generation: Simple algorithm.

Step 1.	$R := \emptyset$
Step 2.	L be the set of supported itemsets with cardinality larger than 2 ($thr - s$)
Step 3.	for all $l \in L$
Step 4.	for all $X \subset l$ with $X \neq \emptyset$ (generate all possible rules from l)
Step 5.	if ($Confidence(X \Rightarrow (l \setminus X)) \geq thr - c$ then
Step 6.	$R := R \cup (X \Rightarrow (l \setminus X))$
Step 7.	end if
Step 8.	end for
Step 9.	end for
Step 10.	return R

For supported itemsets, the following holds.

Lemma 2.2 *Let I_0 be a supported itemset. Then, any non empty subset I_0' of I_0 (i.e., $I_0' \subseteq I_0$ such that $|I_0'| \geq 1$) is also supported* □

Proof From Lemma 2.1 and the fact that I_0 is supported, it follows

$$Support(I_0') \geq Support(I_0) \geq thr - s.$$

So, I_0' is also supported. □

In addition we will also reject the rules below a certain confidence level. That is, the rules that fail too often. For example, rules that apply correctly less than 75% of the times are not valid. We will denote this threshold by $thr - c$.

So as a summary, we are interested in finding rules R such that

$$Support(R) \geq thr - s \tag{2.14}$$

and

$$Confidence(R) \geq thr - c. \tag{2.15}$$

Algorithms to find such rules are known by rule mining algorithms. Algorithm 4 is a simple algorithm for rule mining. The algorithm first considers all supported itemsets. That is, it selects all itemsets with a support larger than $thr - s$. Then, for each of these itemsets, it generates all possible rules, and all those rules with enough confidence are returned.

The cost of this algorithm is very large. If the number of items in the itemset I is m, there are 2^m subsets of I. Of these, $2^m - m - 1$ are the number of subsets of I with cardinality larger than or equal to 2. With small m, the cost becomes unfeasible. For example for the cardinalities (number of different products in the supermarket) $m = 10, 20,$ and 100 we have $2^{10} = 1024, 2^{20} \approx 10^6,$ and $2^{100} \approx 10^{30}$.

Because of this, more efficient and heuristic methods have been defined to find relevant and interesting rules.

Support and confidence can be understood as probabilities. In particular, the support of an itemset X can be understood as the probability that X occurs. Therefore, we can estimate $P(X)$ as follows:

$$P(X) = Support(X) = \frac{\text{transactions satisfying } X}{\text{number of transactions}}.$$

Then, the confidence of rule $R = (X \Rightarrow Y)$ can be understood as the conditional probability of $X \cup Y$ given X. So, we can estimate the probability for this rule $P(Y|X)$ as

$$P(Y|X) = \frac{P(X \cup Y)}{P(X)}.$$

2.3.2.1 Apriori Algorithm

The Apriori algorithm [46] is a well known algorithm for association rule learning. The algorithm incrementally defines candidate itemsets of length k from itemsets of length $k - 1$. This process is based on the following lemma, which is known as the *downward closure lemma*.

Lemma 2.3 *[47] Let L_k be all itemsets with cardinality k. That is,*

$$L_k = \{S||S| \geq k, Support(K) \geq thr - s\}.$$

Then, if L_k is empty, $L_{k'}$ is empty for all $k' > k$.

Proof Let us presume that L_k is empty but $L_{k'}$ is not for $k' = k + 1 > k$. This means that there exists a supported itemset I_0 of cardinality $k + 1$. Let i_0 be an item in I_0. Then, $I_0 \setminus \{i_0\}$ is also supported by Lemma 2.2. As I_0 has cardinality k, we have a contradiction and the proposition is proved. $\qquad\square$

This result permits us to define an algorithm that considers supported itemsets of increasing cardinality. For each itemset L_{k-1} of cardinality $k - 1$ we will define a candidate set for the itemsets of cardinality k and then prune all those that are not sufficiently supported. The supported ones will define L_k. Algorithm 5 describes this process.

The process of constructing the new candidate set is given in Algorithm 6 (see [46,48]). For each pair of itemsets J_1 and J_2 that share all items except one we compute the union that will have exactly k items. The union will be in C_k. Formally,

$$C_k = \{J_1 \cup J_2 | J_1, J_2 \in L_{k-1} \text{ and } |J_1 \cap J_2| = k - 2\}.$$

Algorithm 5: Association Rule Generation: Apriori algorithm.

Step 1.	L_1 be the set of supported itemsets of cardinality one$(thr - s)$.
Step 2.	Set $k := 2$
Step 3.	while $(L_{k-1} \neq \emptyset)$
Step 4.	$C_k :=$ new candidate set (L_{k-1})
Step 5.	$L_k :=$ remove non supported itemsets in $C_k(thr - s)$
Step 6.	$k := k + 1$
Step 7.	end while
Step 8.	return $L_1 \cup L_2 \cup \cdots \cup L_k$

Algorithm 6: Association Rule Generation: Apriori algorithm. Computation of the new candidate set from L_{k-1}. If the elements in the itemsets J_1 and J_2 are ordered, this order can be exploited to speed up the procedure.

Step 1.	$C_k := \emptyset$
Step 2.	for each pair J_1, J_2 in L_{k-1}
Step 3.	if $(J_1$ and J_2 share $k - 2$ items) then
Step 4.	$C_k := C_k \cup \{J_1 \cup J_2\}$
Step 5.	end if
Step 6.	end for
Step 7.	return C_k

For example, let us consider $k = 5$ and $L_{k-1} = L_4$ including, among others, $J_1 = \{a, b, c, d\}$ and $J_2 = \{a, b, c, e\}$. Then, as J_1 and J_2 share $k - 2 = 5 - 2 = 3$ items, we will include in C_5 the itemset $J_1 \cup J_2 = \{a, b, c, d, e\}$.

In order to know if an itemset $c \in C_k$ should be in L_k or not, we first check whether its subsets of cardinality $k - 1$ are in L_{k-1}. If one fails to be in L_{k-1}, then c should not be in L_k. However, this is not enough to ensure that the support of c is larger than the threshold. This has to be checked also. Algorithm 7 describes this process.

In the remaining part of this section we consider the application of the Apriori algorithm (Algorithm 5) to the database in Table 2.2. This example is from [46]. We will use a threshold of $thr - s = 2/4$. For the sake of simplicity we will work with the count instead of the support, so, we will use a threshold of 2.

The algorithm starts with the definition of supported itemsets of cardinality one. These itemsets define L_1. Using the database in Table 2.2, we compute the following count values:

- Count($\{1\}$) = 2
- Count($\{2\}$) = 3
- Count($\{3\}$) = 3
- Count($\{4\}$) = 1
- Count($\{5\}$) = 3

Algorithm 7: Association Rule Generation: Apriori algorithm. Removal of non supported itemsets in $C_k(thr - s)$ to compute L_k.

Step 1.	for all c in C_k
Step 2.	for all subsets c' of c with $k - 1$ elements
Step 3.	remove c from C_k if c' is not in L_{k-1}
Step 4.	end for
Step 5.	end for
Step 6.	for all c in C_k
Step 7.	if $(Support(c) < thr - s)$ then remove c from C_k
Step 8.	end for
Step 9.	return C_k

Table 2.2 Database for the example of the Apriori algorithm, from [46]

Transaction number	Itemsets
x_1	$\{1, 3, 4\}$
x_2	$\{2, 3, 5\}$
x_3	$\{1, 2, 3, 5\}$
x_4	$\{2, 5\}$

Therefore, L_1 consists of all items except 4. That is,

$$L_1 = \{\{1\}, \{2\}, \{3\}, \{5\}\}.$$

The next step (Step 4) is to compute C_2, the candidate set of itemsets with cardinality 2. This is computed from L_1 using Algorithm 6. This consists in combining itemsets from L_1 such that they have all elements except one in common. In the case of C_2 this corresponds to the pairs J_1 and J_2 from L_1. Therefore, we get

$$C_2 = \{\{1, 2\}, \{1, 3\}, \{1, 5\}, \{2, 3\}, \{2, 5\}, \{3, 5\}\}.$$

Let us now apply Step 5, which consists in removing all non supported itemsets from C_2 to define L_2. To do so we apply Algorithm 7. So, first we have to remove all elements in C_2 with subsets not in L_1. In our case, there is no such set as all subsets of itemsets in C_2 are in L_1. Then, we have to check in the database if itemsets in C_2 are supported. The following counts are found:

- $Count(\{1, 2\}) = 1$
- $Count(\{1, 3\}) = 2$
- $Count(\{1, 5\}) = 1$
- $Count(\{2, 3\}) = 2$
- $Count(\{2, 5\}) = 3$
- $Count(\{3, 5\}) = 2$

As $thr - s = 2$, we have

$$L_2 = \{\{1, 3\}, \{2, 3\}, \{2, 5\}, \{3, 5\}\}.$$

From this set, we compute C_3 obtaining

$$C_3 = \{\{1, 2, 3\}, \{1, 2, 5\}, \{1, 3, 5\}, \{2, 3, 5\}\}.$$

Then, L_3 will contain only the itemsets from C_3 that are supported using Algorithm 7. Step 3 in this algorithm removes $\{1, 2, 3\}$ and $\{1, 2, 5\}$ as $\{1, 2\}$ is not supported, $\{1, 3, 5\}$ as $\{1, 5\}$ is not supported and only $\{2, 3, 5\}$ remains. Then, in Step 7 we check in the database if $\{2, 3, 5\}$ is supported and as this is so ($Count(\{2, 3, 5\}) = 2$) we get

$$L_3 = \{2, 3, 5\}.$$

Then, in the next step we compute C_4, but this is empty, so the algorithm finishes and we have obtained the following sets:

- $L_1 = \{\{1\}, \{2\}, \{3\}, \{5\}\}$,
- $L_2 = \{\{1, 3\}, \{2, 3\}, \{2, 5\}, \{3, 5\}\}$,
- $L_3 = \{\{2, 3, 5\}\}$.

Therefore, the algorithm returns these sets from which rules will be generated.

2.3.3 Expectation-Maximization Algorithm

In this section we describe the EM algorithm, where EM stands for expectation-maximization. The algorithm looks for maximum likelihood estimates. We define in this section first likelihood estimate and then the EM algorithm.

2.3.3.1 Likelihood Function and Maximum Likelihood

The *maximum likelihood* is a method for estimating the parameters of a given probability density. Let us consider a probability density $f(z|\theta)$. That is, f is a parametric model of the random variable z with parameter θ (or, parameters, because θ can be a vector). Let $\mathbf{z} = \{z_1, \ldots, z_e\}$ be a sample of the variable z. Then, the *likelihood* of z under a particular model $f(z|\theta)$ is expressed by:

$$f(\mathbf{z} = (z_1, \ldots, z_e)|\theta) = \prod_{i=1}^{e} f(z_i|\theta)$$

That is, $f(\mathbf{z}|\theta)$ is the probability of the sample $\mathbf{z}$ under the particular model $f(z_i|\theta)$ with a particular parameter θ. The likelihood function is the function above when the sample is taken as constant and θ is the variable. This is denoted by $L(\theta|\mathbf{z})$. Thus,

$$L(\theta|\mathbf{z}) = \prod_{i=1}^{e} f(z_i|\theta)$$

Often, the *log-likelihood* function is used instead of the likelihood function. The former is the logarithm of the latter and is denoted by $l(\theta|\mathbf{z})$ or, sometimes, by $l(\theta)$. Therefore,

$$l(\theta|\mathbf{z}) = \log L(\theta|\mathbf{z}) = \log \prod_{i=1}^{e} f(z_i|\theta) = \sum_{i=1}^{e} \log f(z_i|\theta)$$

Given a sample $\mathbf{z}$ and a model $f(\mathbf{z}|\theta)$, the maximum likelihood estimate of the parameter θ is the $\hat{\theta}$ that maximizes $l(\theta|\mathbf{z})$. Equivalently, the estimate is $\hat{\theta}$ such that

$$l(\theta|\mathbf{z}) \le l(\hat{\theta}|\mathbf{z})$$

for all θ.

2.3.3.2   EM Algorithm

The EM algorithm [49] (where EM stands for *Expectation-Maximization*) is an iterative process for the computation of maximum likelihood estimates. The method starts with an initial estimation of the parameters and then in a sequence of two step iterations builds more accurate estimations. The two steps considered are the so-called Expectation step and Maximization step.

The algorithm is based on the consideration of two sample spaces $\mathscr{Y}$ and $\mathscr{X}$ and a many-to-one mapping from $\mathscr{X}$ to $\mathscr{Y}$. We use y to denote this mapping, and $X(y)$ to denote the set $\{x|y = y(x)\}$. Only data y in $\mathscr{Y}$ are observed, and data x in $\mathscr{X}$ are only observed indirectly through y. Due to this, x are referred to as complete data and y as the observed data.

Let $f(x|\theta)$ be a family of sampling densities for x with parameter θ, it is clear that the corresponding family of sampling densities $g(y|\theta)$ can be computed from $f(x|\theta)$ as follows:

$$g(y|\theta) = \int_{X(y)} f(x|\theta)dx$$

Now, roughly speaking, the expectation step consists on estimating the complete data x and the maximization step consists on finding a new estimation of the parameters θ by maximum likelihood. In this way, the EM algorithm tries to find the value θ that maximizes $g(y|\theta)$ given an observation y. However, the method also uses $f(x|\theta)$.

References

1. Babbage, C.: Passages from the life of a philosopher, Longman, Green, Longman, Roberts & Green (1864)
2. Breiman, L.: Statistical modeling: the two cultures. Stat. Sci. **16**(3), 199–231 (2001)
3. Friedman, J.H.: Data mining and statistics: what's the connection? (1997). http://www-stat.stanford.edu/~jhf/ftp/dm-stat.pdf
4. http://brenocon.com/blog/2008/12/statistics-vs-machine-learning-fight/. Accessed Jan 2017
5. http://normaldeviate.wordpress.com/2012/06/12/statistics-versus-machine-learning-5-2/. Accessed Jan 2017
6. http://stats.stackexchange.com/questions/1521/data-mining-and-statistical-analysis. Accessed Jan 2017
7. http://stats.stackexchange.com/questions/6/the-two-cultures-statistics-vs-machine-learning. Accessed Jan 2017
8. Geisser, S.: Predictive Inference: An Introduction. CRC Press (1993)
9. Srivastava, M.S.: Minimum distance classification rules for high dimensional data. J. Multivar. Anal. **97**(9), 2057–2070 (2006)
10. Yata, K., Aoshima, M.: Effective PCA for high-dimension, low-sample-size data with noise reduction via geometric representations. J. Multivar. Anal. **105**(1), 193–215 (2012)
11. Yata, K., Aoshima, M.: Correlation tests for high-dimensional data using extended cross-data-matrix methodology. J. Multivar. Anal. **117**, 313–331 (2013)
12. Hastie, T., Tibshirani, R., Friedman, J.: The Elements of Statistical Learning. Springer (2009)
13. Witten, I.H., Frank, E., Hall, M.A.: Data Mining. Elsevier (2011)
14. Ryan, T.P.: Modern Regression Methods. Wiley (1997)
15. Dubes, R., Jain, A.K.: Clustering techniques: the user's dilemma. Pattern Recogn. **8**, 247–260 (1976)
16. Jain, A.K., Dubes, R.C.: Algorithms for Clustering Data. Prentice Hall, Englewood Cliffs (1988)
17. Berend, D., Tassa, T.: Improved bounds on bell numbers and on moments of sumes of random variables. Probab. Math. Stat. **30**(2), 185–205 (2010)
18. Kaufman, L., Rousseeuw, P.J.: Finding Groups in Data: An Introduction to Cluster Analysis. Wiley (1990)
19. Torra, V., Endo, Y., Miyamoto, S.: On the comparison of some fuzzy clustering methods for privacy preserving data mining: towards the development of specific information loss measure. Kybernetika **45**(3), 548–560 (2009)
20. Torra, V., Endo, Y., Miyamoto, S.: Computationally intensive parameter selection for clustering algorithms: the case of fuzzy c-means with tolerance. Int. J. Intel. Syst. **26**(4), 313–322 (2011)
21. Zadeh, L.A.: Fuzzy sets. Inf. Control **8**, 338–353 (1965)
22. Bezdek, J.C.: The parable of Zoltan. In: Seising, R., Trillas, E., Moraga, C., Termini, S. (eds.) On Fuzziness: Volume 1 (STUDFUZZ 298), pp. 39–46. Springer (2013)
23. Ruspini, E.H.: A new approach to clustering. Inf. Control **15**, 22–32 (1969)
24. Bezdek, J.C.: Pattern Recognition with Fuzzy Objective Function Algorithms. Plenum Press, New York (1981)
25. Miyamoto, S.: Introduction to Fuzzy Clustering (in Japanese), ed. Morikita, Japan (1999)

26. Miyamoto, S., Ichihashi, H., Honda, K.: Algorithms for Fuzzy Clustering. Springer (2008)
27. Höppner, F., Klawonn, F., Kruse, R., Runkler, T.: Fuzzy Cluster Analysis. Wiley (1999)
28. Miyamoto, S., Mukaidono, M.: Fuzzy c-means as a regularization and maximum entropy approach. In: Proceedings of the 7th International Fuzzy Systems Association World Congress, IFSA 1997, vol. II, pp. 86–92 (1997)
29. Miyamoto, S., Umayahara, K.: Fuzzy c-means with variables for cluster sizes (in Japanese). In: 16th Fuzzy System Symposium, pp. 537–538 (2000)
30. Ichihashi, H., Honda, K., Tani, N.: Gaussian mixture PDF approximation and fuzzy c-means clustering with entropy regularization. In: Proceedings of the 4th Asian Fuzzy System Symposium, 31 May–3 June, Tsukuba, Japan, pp. 217–221 (2000)
31. Davé, R.N.: Characterization and detection of noise in clustering. Pattern Recogn. Lett. **12**, 657–664 (1991)
32. Krishnapuram, R., Keller, J.M.: A possibilistic approach to clustering. IEEE Trans. Fuzzy Syst. **1**, 98–110 (1993)
33. Barni, M., Cappellini, V., Mecocci, A.: Comments on "a possibilistic approach to clustering". IEEE Trans. Fuzzy Syst. **4**(3), 393–396 (1996)
34. Ladra, S., Torra, V.: On the comparison of generic information loss measures and cluster-specific ones. Int. J. Uncertainty Fuzziness Knowl. Based Syst. **16**(1), 107–120 (2008)
35. Pal, N.R., Pal, K., Bezdek, J.C.: A mixed c-means clustering model. In: Proceedings of the 6th IEEE International Conference on Fuzzy Systems, pp. 11–21 (1997)
36. Pal, N.R., Pal, K., Keller, J.M., Bezdek, J.C.: A possibilistic fuzzy c-means clustering algorithm. IEEE Trans. Fuzzy Syst. **13**(4), 517–530 (2005)
37. Rand, W.M.: Objective criteria for the evaluation of clustering methods. J. Am. Stat. Assoc. **66**(336), 846–850 (1971)
38. Hubert, L.J., Arabie, P.: Comparing partition. J. Classif. **2**, 193–218 (1985)
39. Anderson, D.T., Bezdek, J.C., Popescu, M., Keller, J.M.: Comparing fuzzy, probabilistic, and possibilistic partitions. IEEE Trans. Fuzzy Syst. **18**(5), 906–918 (2010)
40. Albatineh, A.N., Niewiadomska-Bugaj, M., Mihalko, D.: On similarity indices and correction for chance agreement. J. Classif. **23**, 301–313 (2006)
41. Morey, L., Agresti, A.: The measurement of classification agreement: an adjustment to the rand statistic for chance agreement. Educ. Psychol. Meas. **44**, 33–37 (1984)
42. López de Mántaras, R.: A distance-based attribute selection measure for decision tree induction. Mach. Learn. **6**, 81–92 (1991)
43. Hüllermeier, E., Rifqi, M.: A fuzzy variant of the Rand index for comparing clustering structures. In: Proceedings, IFSA/EUSFLAT (2009)
44. Brouwer, R.K.: Extending the Rand, adjusted Rand and jaccard indices to fuzzy partitions. J. Intell. Inf. Syst. **32**, 213–235 (2009)
45. Hüllermeier, E., Rifqi, M., Henzgen, S., Senge, R.: Comparing fuzzy partitions: a generalization of the rand index and related measures. IEEE Trans. Fuzzy Syst. **20**(3), 546–556 (2012)
46. Agrawal, R., Srikant, R.: Fast algorithms for mining association rules. In: Proceedings of the 20th International Conference on VLDB, pp. 478-499. Also as research report RJ 9839, IBM Almaden Research Center, San Jose, California, June 1994
47. Agrawal, R., Imielinski, T., Swami, A.N.: Mining association rules between sets of items in large databases. In: Proceedings of the 1993 ACM SIGMOD International Conference on Management of Data, pp. 207-216 (1993)
48. Mannila, H., Toivonen, H., Verkamo, A.I.: Efficient algorithms association for discovering rules. AAAI technical report WS-94-03. http://www.aaai.org/Papers/Workshops/1994/WS-94-03/WS94-03-016.pdf (1994)
49. Dempster, A.P., Laird, N.M., Rubin, D.B.: Maximum likelihood from incomplete data via the EM algorithm. J. Roy. Stat. Soc. **39**, 1–38 (1977)

On the Classification of Protection Procedures

3

Aquesta mar no s'assembla gens a la nostra. És una llenca metàl.lica, sense transparències, ni colors canviants

C. Riera, Te deix, amor, la mar com a penyora, 2015, p. 11 [1]

The literature on data privacy is vast, and nowadays a large number of procedures have been developed to ensure data privacy. Different methods have been developed for different scenarios and under different assumptions on the data. In this chapter we describe three dimensions that can be used to classify some of the existing procedures and place some of the existing methods within this classification. The first section describes the dimensions. In the remaining sections we focus on each of these dimensions and discuss particular classes of methods.

3.1 Dimensions

We consider methods for data privacy according to

- whose privacy is being sought,
- the computations to be done, and
- the number of data sources.

Individuals and organizations play different roles with respect to information processing. The first dimension focuses on the individual or organization whose

© Springer International Publishing AG 2017
V. Torra, *Data Privacy: Foundations, New Developments and the Big Data Challenge*, Studies in Big Data 28, DOI 10.1007/978-3-319-57358-8_3

data has to be protected. The second on the type of computation a researcher or a data analyzer needs to apply to the data. From the data protection perspective, it is different if the type of is is known or is ill-defined. The third dimension considers whether the data analyser will use one or more data sets. We describe next these three dimensions in more detail, and discuss some of their relationships.

We also discuss the need for knowledge intensive tools in data privacy. We call *knowledge intensive data privacy* the area that encompasses tools and methodologies where knowledge rich tools play a role. Knowledge intensive data privacy is independent of the previous dimensions and can play a role in several different scenarios.

3.1.1   On Whose Privacy Is Being Sought

This dimension, which was initially discussed in [2], can be illustrated considering a scenario that involves three actors. Then, we can observe privacy from the point of view of each of the actors. We consider these actors: the data respondents, the data holder and the data user.

Literature also uses the term data owner. However, this term is used differently by different authors. For example, [2–4] use it as equivalent to data holder, while [5] uses it as equivalent to data respondent. We will mainly use data holder, but we consider data owner and data holder as equivalents.

The data respondents are the individuals that have generated the data. We consider them as passive subjects that cannot take actions to protect their own privacy. We can consider as typical scenarios the case of data given by people (the respondents) through e.g. research questionaries, interviews, purchases with fidelity cards, medical treatments in hospitals, and information supplied to insurance companies.

The holder is the organization or individual that has gathered the data and has it in a proprietary database. For example, supermarkets have the listings from fidelity cards that describe all items bought by clients, insurance companies all *relevant* data about their clients, and hospitals have the clinical information of their patients, the results of their analyses, and the treatments received. National Statistical Offices have databases with the results of surveys and census.

A user is a person using an information technology service. As in most of the cases, user's access is recorded and stored, we can also consider users as respondents. The main difference between a respondent and a user is that the latter has an active role in order to protect his own privacy. For example, when Alice is accessing a search engine, all her queries will be typically logged into a database together with some additional information as IP address, time stamps, and so on. Because of that, Alice is a respondent of the database with query logs. However, we consider her in the role of a user when she is worried with the privacy of her own queries and she applies herself technological solutions to avoid the disclosure of sensitive information.

Then, we have respondent, holder and user privacy depending on the actor in whose privacy we focus. Before defining these types of privacy below, we summarize the examples we have just discussed.

Example 3.1 A hospital collects data from patients and prepares a server to be used by researchers to explore the data.

The actors in this scenario are the hospital as the data holder, and the patients as respondents whose data is in the database.

When the server logs the queries and computations of the researchers, we have that researchers are respondents of the log database.

Example 3.2 An insurance company collects data from customers for their internal use. They are asking a software company to develop new software. For this purpose a fraction of the database will be transferred to the software company for analysis and test of the new system. The insurance company is the data holder, and the customers are the data respondents.

Example 3.3 A search engine resolves users' queries by sending lists of web pages to these users about the topics queried. Claudia, a user of a search engine, wants to get information about her topics of interest but at the same time avoiding the search engine to know them. She uses a plugin in her browser that has been developed for this purpose. Claudia is a respondent of the search engine database if the engine logs all requests. In addition, Claudia is a user in the sense that she has an active role in order to protect her privacy.

Example 3.4 Two supermarkets have implemented fidelity cards and use them to record all transactions of their customers. The two directors of the supermarket have decided to mine relevant association rules from their databases. In the extent possible, each director does not want the other to access to own records.

In this scenario, the supermarkets are the data holder, and the customers are the respondents.

Example 3.5 Berta, a journalist of a certain country, is sending reports to a foreign newspaper about some activities of her government. The journalist wants to avoid sniffers to know that she has sent the information to the newspaper.

In this case, the journalist is a user of an emailing system and she has an active role to protect her own privacy.

Let us now review the three main categories of privacy under this dimension.

- **Respondents' privacy**. The goal is to avoid disclosure of sensitive data corresponding to particular respondents. In Example 3.1, regulations force the hospital to avoid the disclosure of e.g. illnesses and medical treatments of patients. For example, Elia, a researcher querying the database, should not be able to find that the reason of her friend Donald visiting the hospital is a vasectomy. Due to this, the hospital has to implement respondents' privacy mechanisms to avoid the disclosure of respondents' sensitive information to researchers. Example 3.2 (the case of software development for the insurance company) is also a problem of respondent's privacy.

- **Holder's privacy.** The goal is to avoid the disclosure of any relevant information of the database as this is against the interest of the holder. The scenario in Example 3.4 where two supermarkets collaborate to mine association rules in a way that none of the supermarkets learn anything about the other but the mined association rules corresponds to this type of privacy.
- **User's privacy.** The goal is to ensure the privacy of the user of a particular system, and to achieve this goal, the user acts actively for this purpose. For example, Claudia and Berta in Examples 3.3 and 3.5 apply available technology to avoid as much disclosure of sensitive information as possible. This is also the case of researchers accessing the database in the hospital of Example 3.1 if they try to avoid the hospital to know about their queries.

In this classification, we distinguish between holder's and respondent's privacy. Note that although holders are typically responsible of respondents' privacy according to laws and regulations, holders' interests are not the same as respondents' interests. For example, the director of a hospital may not be really worried that Elia learns about Donald's vasectomy (except for legal liability). The same can apply about Donald's shopping cart in the supermarket. This information is too specific and the owner's focus can be more on generic information and knowledge. For example, the director of the supermarket may be interested in avoiding competitors to know that the pack (*beer, chocolate eggs*) is selling really well, and all people buying them are also buying *chips*. Similarly, the hospital director may be worried that someone learns that most patients of doctor Hyde with appendicitis are back to the hospital after two weeks with a severe illness. In both examples, the information of a particular person may be of no real concern to the data holder.

Methods for user's privacy are discussed in Chap. 4. Methods for respondent and holder privacy are discussed in Sect. 3.2. We include these methods in the same section because although their goal is slightly different, both methods are applied by the data holder.

3.1.2 On the Computations to be Done

This dimension considers the prior knowledge the data holder has about the usage of the data. That is, it focuses on the data use, the expected use of the protected data. Data protection methods can be classified according to their suitability in terms of this available knowledge. For illustration, let us consider two scenarios.

Example 3.6 Aitana, the director of hospital A, contacts Beatriu, the director of hospital B. She proposes to compute a linear regression model to estimate the number of days patients stay in hospital using their databases.

Example 3.7 Elia, a researcher on epidemiology, has contacted Aitana the director of a hospital chain. She wants to access the database because she studies flu and she wants to compare how the illness spreads every year in Chicago and in Miami.

In the first scenario we know the analysis to be computed (a linear regression model) from the data. Privacy can be optimized with respect to this computation. That is, we can develop a privacy approach so that the two directors can compute the linear model and the information leakage is minimal.

In the second scenario, the analysis is ill-defined, and, therefore, we need to apply a data protection method that permits Elia to have good results (as similar as the ones she would obtain with the original data) and where the risk of information leakage is minimal.

In general, the knowledge on the intended data use permits us to distinguish three categories. The first two, data-driven and computation-driven procedures, roughly correspond to the two scenarios above. The third category, result-driven procedures, is loosely related to computation-driven.

We add a third example into this section to illustrate result-driven procedures.

Example 3.8 A retailer specialized in baby goods publishes a database with the information gathered from customers with their fidelity card. This database is to be used by a data miner to extract some association rules. The retailer is very much concerned about alcohol consumption and wants to avoid the data miner inferring rules about baby diapers and beers.[1]

Let us now review the categories.

- **Computation-driven or specific purpose protection procedures**. In this case it is known beforehand the analysis to be applied to the data. Because of that we can develop or use a protection procedure that takes advantage of this fact. In other words, protection procedures are tailored for a specific purpose. We call this type of methods computation-driven as this known computation defines the method or leads to its selection.

  Example 3.6 where Aitana and Beatriu agree on computing a linear regression using the data of the two hospitals corresponds to this case. We are also in this framework when someone, different to the directors, wants to compute the mean of a database.
- **Data-driven or general purpose protection procedures**. In this case, ill-defined or no specific analysis is foreseen for the data. This is the case of Example 3.7. This is also the case when data is published in a web server for public use, as some governmental offices or research organizations do. See e.g. information from a National Statistical Office in [7], and the UCI repository in [8].

---

[1] A classic example in the literature of association rule mining is about the discovery of a rule stating that men that buy diapers also buy beers (see e.g. [6]).

As the intended use is ill defined, we just transfer data to researchers (a third party) so that they can compute their analysis. However, in order to avoid disclosure risk we cannot send the original data but some data that is *similar* to the original one. Similar in the sense that the results and inferences obtained from these data are the same (or similar) to the ones researchers would have obtained from the original data.

The quality requirements for the protected data may depend on the type of scenario. In Example 3.2, the quality of the analysis can be low if our only purpose is that software engineers can test their software (although in this case data needs to have enough coverage) as there is no need to analyse the output results. In contrast, when a researcher wants to study the effectiveness of a medical treatment, inferences should be correct.

An intermediate case is when data is generated for educational purposes (see e.g. [9, p. 31] "to train students in the application of multivariate statistical techniques on 'real' data").

- **Result-driven protection procedures**. In this case, privacy concerns to the result of applying a particular data mining method to some particular data [10, 11]. Example 3.8 illustrates this case.

  Although data protection methods preventing this type of disclosure are specific to a given computation procedure, we classify them in a specific class because the focus is different. We focus here on the knowledge inferred from the data instead of the data itself.

Note that the dimension discussed in this section mainly focuses on respondent's and holders's privacy, and that methods to ensure privacy for them are developed to be applied by the data holder.

3.1.3 On the Number of Data Sources

The last dimension we discuss corresponds to the number of data sources. We distinguish only two cases.

- **Single data source**. That is, we only consider one data set. For example, the data analysis is based on a single database. Most examples discussed above correspond to this case.
- **Multiple data sources**. In this case, different data sets are under consideration. We mainly find multiple data sources in two types of scenarios.

  On the one hand, we have multiple data sources in the case of analysis of databases belonging to different data holders. Typically, the analysis is a given function. This is the case of Example 3.6, where the databases of the two hospitals are used to build the linear regression model.

  On the other hand, we also have multiple databases when we have multiple releases of the same data set, or releases of different datasets that share variables. In this case, data protection has to take into account that adversaries can integrate these data sets in order to improve their attacks into the original data.

The first scenario of a known function usually falls into the computation-driven approach. In fact, it is in the other way round. Most of the research on computation-driven approaches focus in the case of several datasets. We will discuss this issue in Sect. 3.4. In the more general case in which the function is ill-defined, data-driven approaches are applied. An example of an approach for multiple data sources that is not computation-driven is found in [12].

The second scenario with multiple releases is of relevance in the analysis of disclosure risk. We have also this situation in the case of stream data which requires multiple releases of related data sets. We will discuss it in Sect. 5.9.3.

3.1.4 Knowledge Intensive Data Privacy

Uholak eraman du hitzen eta gauzen arteko zubia

K. Uribe, Bitartean heldu eskutik, 2001, p. 65 [13]

Initial results in both statistical disclosure control and privacy preserving data mining were on files with numerical and categorical data. Later this evolved in more complex data types (logs, social networks, documents). In the last years there has been a trend to deal not only with the data in a database but also with additional information and knowledge on the data, data schema and even on the semantics attached to the data. Knowledge intensive data privacy focuses on these questions.

We outline here some of the aspects that can be considered as part of knowledge intensive data privacy [14].

- **Semantics of terms**. In both data protection and disclosure risk the semantics of the terms present in the file can be taken into account. Initial work in this area includes generalization approaches where the generalization schema is known (e.g., recoding cities by counties, and counties by provinces/states). Knowledge rich methods consider ontologies, dictionaries and language technologies for data protection (e.g., document sanitization) and for risk assesment by means of record linkage.
- **Metadata and constrained data**. Files and tables of databases are not isolated entities, the same for the variables in a file or table. It is common to have relationships (constraints) between the variables and these relationships have to be taken into account in the data protection process. We discuss this problem in Sect. 6.5.
- **Knowledge rich disclosure risk assessment**. Disclosure risk has to be based on the best technologies for database integration and matching. However, not all information is in a structured way, because there is information in noSQL databases, and in free text form (texts in blogs and online social networks). Disclosure risk assessment may need the analysis of such data. The analysis of these unstructured data needs knowledge intensive as well as more traditional tools.

3.1.5 Other Dimensions and Discussion

The literature considers other ways to classify data protection mechanisms. One of them is on the type of tools used. In such dimension, methods are classified either as following the perturbative or the cryptographic approach. Naturally, perturbative methods are the ones that ensure protection by means of adding some type of noise into the original dataset. On the contrary, cryptographic approaches ensure privacy by means of cryptographic protocols.

Cryptographic approaches are typically developed on the basis that we know which is the analysis to be computed from the data. In most of the cases, a cryptographic protocol is defined to compute a particular function of one or more datasets. As the function to be computed is known, they can be classified as computation-driven protection methods. They are also extensively used for user privacy. We will see some of these tools in Chap. 4.

Perturbative approaches are usually considered as independent of the data use, or, in general, they can be typically applied for different data uses. Because of that, most perturbative approaches can be considered as data-driven approaches. As an example of research focused on a specific use (association rules) for a perturbative method (noise addition) see e.g. [10].

We want to underline that in this book we will not use the term *perturbative approach* with this meaning. Instead, we use perturbative methods/approaches in a more restricted way, as it is usual in the statistical disclosure control community. As we show in Chap. 6, we distinguish between perturbative and non-perturbative approaches. Although both reduce the quality of the data, only the first type introduces errors in the data. We will use *masking methods* as the term that encompasses both classes and corresponds to the perturbative approach with the sense above. As a summary, masking methods take the original dataset and construct the protected one reducing the quality of the original one. We will use anonymization methods with a similar meaning.

3.1.5.1 Cryptographic Versus Masking Approaches

The most important differences between a cryptographic approach and a masking approach are that

- cryptographic approaches compute a desired function or analysis exactly, while masking approaches usually compute approximations of the desired function,
- cryptographic approaches ensure a 100% level of privacy, while the masking approach usually does not.

Nevertheless, the better performance of the cryptographic approach with respect to the function to be computed and the optimal level of risk is at the cost of not having flexibility. In order to define a cryptographic protocol to compute a desired analysis, we need to know with detail the analysis to be computed. If the analysis is changed, the whole protocol needs to be redefined. This is not the case for masked data. We can use these data for a variety of purposes, although in most of the cases only approximate results will be obtained. At the same time, masking methods lead to data that may have some disclosure risk.

Another disadvantage of the cryptographic approach, from a computational point of view, is that it is more costly (because of the use of cryptographic protocols) than using a masking approach. Properly speaking, some masking methods are also costly (because optimal solutions are NP-Hard) but it is usual to use heuristic approaches with less cost.

As we will see in Chap. 6, masking methods try to find a good balance between information loss (the quality of the data for analysis) and disclosure risk (in what extend masked data can lead to disclosure). As a summary, we can underline the following:

- **Cryptographic approach**. 100% privacy, 100% accuracy, no flexibility and high computational cost.
- **Masking methods**. Trade-off between privacy and accuracy, low computational cost.

So, each approach has its own advantages and shortcomings. As usual the selection of a method has to be done according to the requirements of the problem.

3.1.5.2 Syntactic and Semantic Methods

The literature on security distinguishes between perfect secrecy and semantic security. Perfect secrecy ensures that a ciphertext does not provide any information without knowing the key. In semantic security it is possible in theory to get information from the ciphertext, but it is not computationally feasible to get this information.

This distinction has led to distinguish between syntactic and semantic methods in data privacy. k-Anonymity is considered a syntactic [15, 16] method while differential privacy is considered an algorithmic and semantic method [16]. From this perspective, computational anonymity as discussed in Sect. 5.8.3 can also be considered as a semantic method (although it is based on k-anonymity).

We will not use this distinction in this book. We will use the term semantic anonymization and e.g. semantic microaggregation when we deal with linguistic terms and take into account their semantics from the point of view of ontologies.

3.1.6 Summary

Figure 3.1 represents the three dimensions we have studied in a graphical way. The main classification is on the dimension of whose privacy is being sought and on who applies the method. Respondent and holder privacy is considered together because methods will be applied by the data holder. In this case, we include the dimension of the type of computation (data-driven, computation-driven and result-driven).

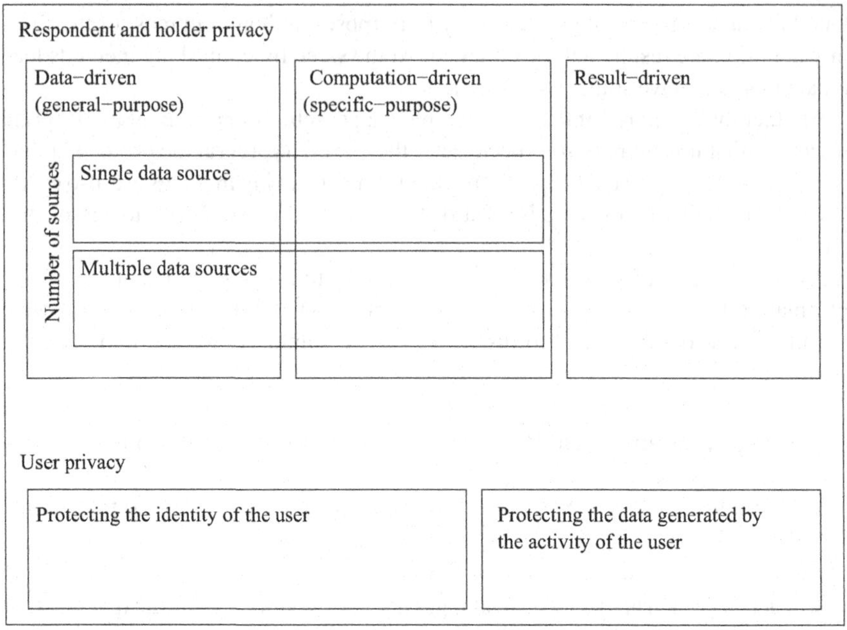

Fig. 3.1 Representation of the three dimensions of data protection methods

Data-driven approaches are mainly used for respondent privacy and computation-driven approaches for holder privacy. Result-driven approaches are mainly applied to ensure holder privacy. Then, we include the third dimension on the number of data sources. This is mainly relevant for data-driven and computation-driven approaches.

Within the user privacy approaches we shall distinguish two types of scenarios that will be discussed later: the protection of the identity of the user and the protection of the data generated by the user. See Chap. 4 for details. Figure 3.1 already shows this distinction.

3.2 Respondent and Holder Privacy

Respondents have no control (except for their indirect influence in legislation) on the level of protection the data holder applies to the data. As explained above, methods for respondent and holder privacy are both applied by the data holder. Because of that, we summarize in this section the main type of scenarios data holders can face with respect to their own privacy protection and the protection of respondents privacy. This is based on our previous discussion.

- **Data-driven methods from a single or from multiple databases**. Different types of masking methods have been developed for this type of scenario. Information loss and disclosure risk measures have been developed to evaluate how good are masking methods. We will outline these methods and measures including references to appropriate sections and chapters in Sect. 3.3. These methods are the ones described in more detail in this book.
- **Computation-driven methods with several data sources**. Such methods are typical for holder privacy and are the typical scenario of cryptographic approaches. We give an outline of these methods in Sect. 3.4.2.
- **Computation-driven methods for a single database release**. If the function is completely specified, this mainly corresponds to querying a database for a particular query. Differential privacy [17] focuses on this scenario. We discuss this topic in Sect. 3.4.1. There is some research in the case of functions not completely specified. In this case the problem is solved with standard data-driven methods taking into account the information on the data use. For example, we can consider the robustness of several data-driven methods when we know that data will be clustered. We will discuss this topic in Sect. 7.4.3.
- **Result-driven methods**. Such methods are mainly used to ensure holder privacy, but they can also be used to avoid discriminatory knowledge inferred from databases. We review these methods in Sect. 3.5.

3.3 Data-Driven Methods

These are the methods that are developed taking into account the type of information to be masked and mainly ignoring the type of analysis to be applied to the data. They can be seen as general purpose data protection methods. In this book we will focus on regular files and databases (i.e., SQL databases), where each row corresponds to data from an individual or entity. This type of data is known as microdata in the statistical disclosure control (SDC) community. We will discuss in Chap. 6 the methods to protect the data. Disclosure risk and information loss measures for regular files is discussed in detail in Chaps. 5 and 7, respectively.

Standard databases is not the only type of data one can encounter in the real world. Another type of data is when we have aggregates of these SQL databases. This is known as tabular data within the SDC community. We will briefly outline the difficulties of data privacy for this type of data in Sect. 3.6 and discuss some of the solutions we find in the literature.

Nowadays there is a large number of databases that do not fit exactly within the more standard framework of *standard* databases. For example, we have document and textual data, and all kinds of big data.

There is no standard definition of big data. Nevertheless, most definitions seem to agree on some basic characteristics (the well-known 3, 4, and 5 V's of big data): volume (large size of data sets), variety (different types of data including text and images), velocity (new data is produced very quickly), variability (data is typically inconsistent), and veracity (the quality of the data can be low).

If we combine volume and velocity, we can classify masking methods according to the following three categories. We will use these categories in Chap. 6 when presenting the different families of data masking procedures, and use them again when we summarize masking methods for big data in Sect. 6.6.

- **(i) Large volumes**. This corresponds to data of high dimension, but with low velocity. Then, masking methods are for *static* (not changing) databases but of huge dimension.
- **(ii) Streaming data**. Data arrives continuously, and needs to be processed in real-time. Data protection methods are usually applied to a sliding window, as it is unfeasible to process all data at once.
- **(iii) Dynamic data**. We have this type of data when a database changes with respect to time and we want to publish different copies of this database. Each release needs to take into account previous releases, and how they have been created, otherwise disclosure can take place.

3.4 Computation-Driven Methods

We divide this section in two parts. First we discuss the case of computation-driven from a single database, and then the case of multiple databases.

For a single database we focus on differential privacy. For multiple databases we focus on cryptographic approaches.

A related approach is homomorphic encryption [18] (e.g., ElGamal cryptosystem). Recall that with homomorphic encryption we can operate on encrypted data and then decrypt the final result. So, for instance, we can use homomorphic encryption when our database is in the cloud so that the cloud owner does not have access to our sensitive data. This approach is related to the cryptographic approaches for multiple databases. Note that as homomorphic encryption can be used for several types of computations, we can classify it within data-driven methods.

3.4.1 Single Database: Differential Privacy

A single computation from a single database can be seen as a query to the database. There are different approaches to study this problem. One of them is query auditing. Following [19], we say that query auditing "is to determine if answering a user's database query could lead to a privacy breach". Reference [5] classifies query auditing into the following two classes.

- **Online auditing**. Analysis is made to each query, and queries that compromize information are not served. For example, [20] studies the case where given a new query, and a set of views from the database already served, the new query is only served when it does only supply information already released. Naturally, when a denial occurs, this gives information to the user.
- **Offline auditing**. Analysis is done once all queries are already made and served. In this case, if there is a privacy breach in the queries, at the time of the analysis the privacy breach has already taken place. Reference [19] studies this type of scenario.

Another approach is to serve the queries but in a way that there is no disclosure of sensitive information. This is the case of differential privacy [17]. In a differential privacy model, queries are served, but the answer includes some error. That is, the answer that we would obtain directly from the database is different from the one obtained by the differential privacy model.

Privacy with respect to queries has also been studied in relation to multilevel secure databases. They are databases in which different levels of security clearances are granted to different users. Reference [21] discusses some of the problems that need to be considered. One is about the inference from queries based on sensitive data, when the answers of different queries are combined to infer sensitive information. This is a problem that also appears with standard databases and studied in e.g. differential privacy. Another is about the inference from data combined with metadata.

Prior to differential privacy, a related approach was proposed in [22,23] for particular types of queries. More specifically, a query consisted of a pair (S, f) where S is a set of rows in the database, and f a function. The function was restricted to be from a row to $\{0, 1\}$ in [22] and to $\mathbb{R}$ in [23]. Then, while the correct response of the database is $nr = \sum_{r \in S} f(DB_r)$, a noise version of nr was given to the user to ensure privacy.

Differential privacy [17] (see also [24,25]) is a Boolean definition of privacy specialized on database queries. As we see below it establishes a condition that should be satisfied by a database system in order to fulfill the differential privacy requirement. Being a Boolean definition it implies that algorithms and research focus on algorithms that minimize information loss.

In [17], Dwork discusses "what constitutes a failure to preserve privacy? What is the power of the adversary whose goal it is to compromise privacy?". According to her, the answer comes from the difference of adding or removing an element in the database. Because of that "differential privacy ensures that the removal or addition of a single database item does not (substantially) affect the outcome of any analysis" (see [24, p. 2]). Basically, systems should be constructed in a way that the answer is similar enough when we add or remove a record in the database. The formal definition of differential privacy is in terms of probability distributions.

The definition follows. We state that two databases D_1 and D_2 differ in at most one element if one is a proper subset of the other and the larger database contains just one additional row. Here, q is the query and $K_q(D)$ is the mechanism used to respond to the query for database D.

Algorithm 8: Differential privacy for a numerical response.

Data: D: Database; q: query; ε: parameter of differential privacy;
Result: Answer to the query q satisfying ε-differential privacy
1 $a := q(D)$ with the original data
2 Compute $\Delta_{\mathscr{D}}(q)$, the sensitivity of the query for a space of databases D
3 Generate a random noise r from a $L(0, b)$ where $b = \Delta(q)/\varepsilon$
4 Return $a + r$

Definition 3.1 A function K_q for a query q gives ε-differential privacy if for all data sets D_1 and D_2 differing in at most one element, and all $S \subseteq Range(K_q)$,

$$\frac{Pr[K_q(D_1) \in S]}{Pr[K_q(D_2) \in S]} \le e^{\varepsilon}.$$

Here, ε is the level of privacy required. The smaller the ε, the greater the privacy we have.

An important concept within differential privacy is the one of sensitivity of a given query.

Definition 3.2 [25] Let $\mathscr{D}$ denote the space of all databases; let $q : \mathscr{D} \to \mathbb{R}^d$ be a query; then, the sensitivity of q is defined

$$\Delta_{\mathscr{D}}(q) = \max_{D, D' \in \mathscr{D}} ||q(D) - q(D')||_1.$$

where $|| \cdot ||_1$ is the L_1 norm, that is, $||(a_1, \dots, a_d)||_1 = \sum_i |a_i|$.

Note that this definition is essentially meaningful when data has upper and lower bounds [26,27].

Differential privacy is usually implemented for numerical data adding noise to the true response. The noise usually follows a Laplace distribution with mean equal to zero and parameter $b = \Delta(q)/\varepsilon$. Following [28] the procedure is described in Algorithm 8. Section 6.1.3 discusses additive and multiplicative noise. Reference [29] introduces an alternative approach based on wavelet transforms. Data masking using wavelet transforms is discussed in Sect. 6.1.5.

There is a lot of research on mechanisms so that differential privacy is satisfied. At the same time there is some discussion about the shortcomings of differential privacy for some type of queries. See e.g. [27,28].

In the model above for differential privacy no consideration is given to the previous knowledge that the intruder has. Note that the error r in Algorithm 8 is only based on the $\Delta_{\mathscr{D}}(q)$ and the distribution selected. An approach for differential privacy considering previous knowledge is described in [30].

Differential privacy has been used for different types of data, not only numerical. In particular, to data of large dimensions. See e.g., [31] for its application to search logs, and [32,33] for its application to graphs (for social networks).

3.4.2 Multiple Databases: Cryptographic Approaches

In this scenario we know the type of analysis, a function, to be computed from the data. The computation will use several databases. Typically, each database has a holder (a party), and the goal is to develop a system so that the function is computed in such a way that the only new information obtained by each of the data holders is the result of the analysis itself. That is, no extra knowledge is acquired in the computation. This is illustrated in the following example.

Example 3.9 Parties $P_1, \ldots, P_n$ own databases $DB_1, \ldots, DB_n$. The parties want to compute a function, say f, of these databases (i.e., $f(DB_1, \ldots, DB_n)$) without revealing unnecessary information. In other words, after computing $f(DB_1, \ldots, DB_n)$ and delivering this result to all P_i, what P_i knows is nothing more than what can be deduced from DB_i and the function f. So, the computation of f has not given P_i any extra knowledge.

A *trivial* approach for solving this problem is to consider a trusted third party TTP that computes the analysis. This is known as the centralized approach. In this case, each P_i transfers data DB_i using a completely secure channel (e.g., using cryptographic protocols) to the trusted third party TTP. Then, TTP computes the result, and sends it to each P_i.

Distributed privacy preserving data mining studies how to build approaches where this trusted third party is not needed, and where the function is computed in a collaborative manner by the parties P_i. Cryptographic protocols are also used for this purpose. The ultimate goal is that the information gain of each data holder is just the result of the function. So, the centralized approach is considered as a reference with respect to disclosure risk for distributed approaches.

Distributed privacy preserving data mining is based on the secure multiparty computation, which was introduced by Yao in 1982 [34]. For example, [35, 36] defined a method based on cryptographic tools for computing a decision tree from two data sets owned by two different parties. Reference [37] discusses clustering data from different parties. See also the book by Vaidya et al. [38] for a detailed description of a few methods following this approach. Recall that all these protocols need to be defined for a particular function f. So, the literature in this area is vast as results are specific for different types of data mining, machine learning and statistical methods.

When data is represented in terms of records and attributes, two typical scenarios are considered in the literature: vertical partitioning of the data and horizontal partitioning. They are as follows.

- **Vertically partitioned data**. All data holders share the same records, but different data holders have information about different attributes (i.e., different data holders have different views of the same records or individuals).

- **Horizontally partitioned data**. All data holders have information about the same attributes, nevertheless the records or individuals included in their databases are different.

As stated above, for both centralized and distributed approaches the only information that should be learnt by the data holders is the one that can be inferred from their original data and the final computed analysis. The centralized approach is considered as a reference result when analyzing the privacy of the distributed approach. Privacy leakage for the distributed approach is usually analyzed considering two types of adversaries.

- **Semi-honest adversaries**. Data holders follow the cryptographic protocol but they analyse all the information they get during its execution to discover as much information as they can.
- **Malicious adversaries**. Data holders try to fool the protocol (e.g. aborting it or sending incorrect messages on purpose) in order to infer confidential information.

3.4.3 Discussion

We finish this section with a discussion on the advantages and disadvantages of computation-driven protection methods. Part of this discussion reproduces what was stated in Sect. 3.1.5.

The use of computation-driven protection procedures based on cryptography present some clear advantages with respect to general purpose ones. The first one is the good quality of the computed function (analysis). That is, the function we compute is exactly the one the users want to compute. This is not so, as we will see later, when other general purpose protection methods are used. In this latter case, the resulting function is just an approximation of the function we would compute from the original data. At the same time, cryptographic tools ensure an optimal level of privacy.

Nevertheless, this approach has some limitations. The first one is that we need to know beforehand the function (or analysis) to be computed. As different functions lead to different cryptographic protocols, any change on the function to be computed (even small ones) requires a redefinition of the protocol. A second disadvantage is that the computational costs of the protocols are very high. In addition, it is even harder when malicious adversaries are considered.

Reference [39] discusses other limitations. One is that most literature only considers the types of adversaries described above (honest, semi-honest and malicious). No other types are studied. Another one is the fact that in these methods no trade-off can be found between privacy and information loss (they use the term accuracy). As we will see later in Sects. 7 and 8, most general purpose protection procedures permit the user to select an appropriate trade-off between these two contradictory issues. When using cryptographic protocols, the only trade-off that can be implemented easily is the one between privacy and efficiency.

3.5 Result-Driven Approaches

The goal of result-driven protection procedures is to ensure that the results of a data mining process do not lead to disclosure of sensitive information. So, if we publish a protected data set, we want to avoid that researchers applying a certain data mining method to this dataset obtain some undesired outcome.

One application of this type of approaches is for holder privacy, when the holder wants to hide a certain knowledge that can be used by competitors. Another application is to avoid the mining of discriminatory knowledge. See e.g. [40] for details.

The case of result privacy was already illustrated in Example 3.8 where a supermarket wants to avoid the disclosure of rules relating the consumption of *baby diapers* and *beers*. We also discussed the case of a holder avoiding competitors to know that people buying the pack (*beer, chocolate eggs*) is also buying *chips*.

With respect to discrimination, as [41] states, one may tend to think that the application of data mining algorithms to raw data, and the posterior use of the inferred knowledge in decisions is free from discrimination bias. The automatic application of both data mining algorithms and decision support systems gives this impression. Nevertheless, subtle (and not so subtle) bias in the data can cause the inference of discriminatory rules. Let us consider the following example based on [40,41].

Example 3.10 We have a dataset containing the attribute race in which most of the people of a certain race had delay in returning previous credits (or failed with them). However, although the ultimate cause can be low salary, data mining algorithms may use race in classification rules and, therefore, defining discriminatory knowledge.

The case of race is of direct discrimination because a sensitive variable is used in the rule. We will have indirect discrimination if the database contains the variable Zip code and this is correlated with race. Indirect discriminatory knowledge can be inferred by a data mining algorithm and lead to a rule denying credits for people living in a particular ZIP code. Some result-driven algorithms can be used to avoid data mining algorithms to infer such discriminatory rules. See [40] for details.

Result-driven methods are also known by *anonymity preserving pattern discovery* [11], *result privacy* [42], *output secrecy* [43], and *knowledge hiding* [44].

In this section we present some examples about result-driven approaches focusing on the case of itemsets and association rules. Let us first consider a formalization of the problem.

Definition 3.3 [44] Let $\mathscr{D}$ be a database, let A be a parametric data mining algorithm. Then, A run with parameter set Θ is said to have ability to derive knowledge K from $\mathscr{D}$ if and only if K appears either directly in the output of the algorithm or by reasoning from the output. This is expressed by $(A, \mathscr{D}, \Theta) \vdash K$.

Then, K is said to be derivable from $\mathscr{D}$, if there exists any algorithm A with parameter Θ such that $(A, \mathscr{D}, \Theta) \vdash K$.

Given $\mathscr{D}$ the set of derivable knowledge is denoted as $K\,Set_{\mathscr{D}}$. Any knowledge K such that $(A, \mathscr{D}, \Theta) \vdash K$ is in $K\,Set_{\mathscr{D}}$.

Using the concept of derivability our data privacy problem is formulated as follows.

Definition 3.4 Let $\mathscr{D}$ be a database. Let $\mathscr{K} = \{K_1, \ldots, K_n\}$ be the set of sensitive knowledge that must be hidden from $\mathscr{D}$. Then, the problem of hiding knowledge $\mathscr{K}$ from $\mathscr{D}$ consists of transforming $\mathscr{D}$ into a database $\mathscr{D}'$ such that

1. $\mathscr{K} \cap K\,Set_{\mathscr{D}'} = \emptyset$,
2. the information loss from $\mathscr{D}$ to $\mathscr{D}'$ is minimal.

To make this definition concrete we need to formalize what means minimal information loss in our context. Reference [44] expresses information loss in terms of misses (false negative) and fakes (false positive). In our context a false negative is when a certain knowledge that should appear in the result (because it is derivable from $\mathscr{D}$) does not appear in the result from $\mathscr{D}'$. On the contrary, a false positive is when a certain result that is not derivable from $\mathscr{D}$ can be derived from $\mathscr{D}'$.

So, [44] replaces condition 2 above by:

- The number of false negative is minimized. That is, minimize $|K\,Set_{\mathscr{D}} \backslash K\,Set_{\mathscr{D}'}|$. Note that here $A \backslash B$ is the set of elements in A not in B.
- The number of false positive is minimized. That is, minimize $|K\,Set_{\mathscr{D}'} \backslash K\,Set_{\mathscr{D}}|$.

Let us now consider the case in which the knowledge inferred from the database are itemsets, and our goal is to prevent the user to infer itemsets. Recall from Sect. 2.3.2 that rule mining finds rules (see Eqs. 2.14 and 2.15) such that

$$Support(R) \geq thr - s$$

and

$$Confidence(R) \geq thr - c$$

for certain thresholds $thr - s$ and $thr - c$.

Due to these two conditions, there are two main approaches for avoiding the disclosure of rules. See e.g. [45,46]. We can

- **Case 1**. Reduce the support of the rule, or
- **Case 2**. Reduce the confidence of the rule. This can be achieved either

  - **Case 2a**. increasing the support of the antecendent, or
  - **Case 2b**. decreasing the support of the consequent (to decrease the support of the rule without decreasing the one of the antecedent).

In order to increase the support of an antecedent, some transactions partially supporting the antecedent (i.e., supporting some subsets of the antecedent but not the antecedent) will be modified to support the antecedent completely. This implies the addition of some items to a transaction. For example, if we have the rule $\{a, b, c, d\} \Rightarrow \{e\}$, and we have the transaction (a, b, c, e, f) which partially supports the rule because d is not present in the transaction, we can increase the support of this rule adding d into the transaction. In order to decrease the support of a consequent, or of a rule, some items will be removed from a transaction. For example, if we have the same rule as before $\{a, b, c, d\} \Rightarrow \{e\}$ and the transaction (a, b, c, d, e, f) we can reduce the support of this rule removing any of the items (a, b, c, d, e) from the transaction.

Note that decreasing the support of the consequent reduces at the same time the support of the rule and the confidence of the rule. As we will see later, Algorithm 10 proceeds in this way, and because of that checks both the confidence and the support of the rule.

Case 1. Reduce the support of the rule. We follow here the approach described in [10]. We consider first that the modifications of a database $\mathcal{D}$ consist only in the modification of transactions by means of deleting some of their items. Then, we consider that information loss relates to the number of itemsets that were first supported and that after the modification are not.

Definition 3.5 [44] Let $\mathcal{D}$ be a database with transactions on the itemset I. Let $thr - s$ be a disclosure threshold. Let $\mathcal{K} = \{K_1, \dots, K_n\}$ be the set of sensitive itemsets and $\mathcal{A}$ the set of non-sensitive itemsets.

Naturally, for all $K \in \mathcal{K} \cup \mathcal{A}$, it holds $K \in 2^I \setminus \emptyset$. In addition, $Support(K) \geq thr - s$ for $K \in \mathcal{K} \cup \mathcal{A}$.

Then, the problem of hiding itemsets $\mathcal{K}$ from $\mathcal{D}$ consists of transforming $\mathcal{D}$ into a database $\mathcal{D}'$ such that

1. $Support_{\mathcal{D}'}(K) < thr - s$ for all $K \in \mathcal{K}$
2. The number of itemsets K in $\mathcal{A}$ such that $Support_{\mathcal{D}'}(K) < thr - s$ is minimized.

This problem was proven to be NP-hard in [10]. Section 3.4 of [47] considers a generalized version of this problem and prove that it is also NP-hard. Because of that, it is interesting to define heuristic algorithms for association rule hiding. The literature presents several algorithms for this purpose (see e.g [10,46,48]). Reference [48] is a recent survey on this topic.

Now we present the algorithm proposed in [10], one of the first approaches to this problem, which focuses on hiding a single itemset. The algorithm, reproduced in Algorithm 9, begins with the itemset to be hidden, and traverses a graph of itemsets until an itemset with two items is reached. In this traversal the parent with maximum support is selected in each step. Once the set with two items is selected, one of the items in the set is removed from a transaction in the database.

Table 3.1 Transactions of a database

Transaction number	Items
T1	a, b, c, d
T2	a, b, c
T3	a, c, d

Table 3.2 Support for itemsets

Itemsets	Count
a, b	2
a, c	3
a, d	2
b, c	2
c, d	2
a, b, c	2
a, c, d	2

Algorithm 9: Modification of a database of transactions. Algorithm from [9] to avoid the disclosure of an association rule.

Data: HI: itemset to be hidden; D: database
Result: A database where HI is not inferred
1 **while** HI *is not hidden* **do**
2     $HI' := HI$
3   **while** $|HI'| > 2$ **do**
4        $P :=$ subsets of HI' with cardinality $|HI'| - 1$
5        $HI' := \arg\max_{hi \in P} Support(hi)$
6     $T_{HI} :=$ set of transactions in the database that support HI
7     $Ts :=$ transactions in T_{HI} that affects the minimum number of itemsets of cardinality 2
8   Remove one item of HI' that is in Ts
9   Propagate results forward

As this algorithm only decreases the support of itemsets, it does not cause false positives in the set of inferred rules. Because of that information loss is focused on false negatives, and the goal of the algorithm is to avoid that valid association rules are no longer inferred from the database.

To illustrate Algorithm 9, let us consider its application to the database of transactions in Table 3.1.

Example 3.11 Let us consider that we have the transactions in Table 3.1 and we want to hide the itemset $HI = \{a, b, c\}$.

Then, we consider all subsets of HI with cardinality $|HI| - 1$. We have $\{a, b\}$, $\{b, c\}$, and $\{a, c\}$. As $Support(\{a, b\}) = Support(\{b, c\}) = 2$, and $Support(\{a, c\}) = 3$ (see Table 3.2), we select the itemset $HI' = \{a, c\}$.

The next step is to find the set of transactions in the database that support HI. This set of transactions is $\{T1, T2\}$. Naturally, all transactions that support HI support also HI' (recall Lemma 2.1).

We define Ts as the transaction $T1$ or $T2$ that affects the minimum number of itemsets of cardinality 2. As $T1 = \{a, b, c, d\}$ and $T2 = \{a, b, c\}$, T2 affects less itemsets than $T1$.

Finally, we remove one of the items in $HI' = \{a, c\}$ that are in T2. As both items have the same support, we select one of them at random.

Finally, we have to propagate the results forward. That is, recompute the support of the itemsets.

Case 2. Reduce the confidence of the rule by means of decreasing the support of the consequent (case 2b). We follow here Algorithm 1.b in [46] reproduced here as Algorithm 10. It decreases the support of the consequent of a rule. This support is reduced until the confidence or the support of the rule are below the corresponding thresholds.

In the algorithm, the iterative process is repeated $NumIter$ times. This number is computed as the minimum between the number of iterations required to decrease the confidence below the confidence threshold $thr - c$, and the number of iterations required to decrease the support below the threshold. These number of iterations are, respectively, $NumIterConf$ and $NumIterSupp$. Expressions for these values are given below. A proof of these values being correct is given Lemmas 4.1 and 4.2 in [46].

The algorithm starts with a rule HI to be hidden in a database $\mathcal{D}$, the algorithm determines the set of transactions in $\mathcal{D}$ that support HI, and counts the number of such transactions in $\mathcal{D}$. This is tc.

Then, the transactions of T_{HI} are ordered in ascending order according to tc. The smallest transaction is selected $t = T_{HI}[1]$.

In the next step, we select an item in t. We select the item with the minimum impact on the database. This selection is explained in [49] as follows: generate all the $(|Y(HI)| - 1)$-subsets of $Y(HI)$ and choose the first item of the $(|Y(HI)| - 1)$-itemset with the highest support, thus minimizing the impact on the $(|Y(HI)| - 1)$-itemsets.

This item is removed from t. Therefore, as t does no longer support HI, it should also be deleted from T_{HI}. Accordingly, *Support(HI)* and *Conf(HI)* of the rule are recomputed.

Algorithm 10: Modification of a database of transactions. Algorithm 1.b from [45] to avoid the disclosure of an association rule.

Data: $HI = (X \Rightarrow Y)$: rule to be hidden, $\mathscr{D}$: database
Result: A database where HI is not inferred
1 $T_{HI} := \{t \in \mathscr{D} | (X \cup Y) \subset t\}$
2 **forall** $t \in T_{HI}$ **do** ;; count all transactions
3 $\quad\lfloor\ tc[t] := Count(t)$
4 $T_{HI} := \text{sortAscendingAccordingToSize}(T_{HI}, tc)$
5 $NumIterConf := \lceil |D| \cdot \left(\frac{Support(HI)}{thr-c} - Support(antecedent(HI)) \right) \rceil$
6 $NumIterSupp := \lceil |D| \cdot \frac{Support(HI)}{thr-s} \rceil$
7 $NumIter := \min (NumIterConf, NumIterSupp)$
8 **for** $i := 1$ *to* NumIter **do**
9 $\quad\big|\quad t := T_{HI}[1]$
10 $\quad\big|\quad j := \text{chooseItem}(t, consequent(HI))$
11 $\quad\big|\quad$ remove item j from t
12 $\quad\big|\quad Support(HI) := Support(HI) - 1$
13 $\quad\big|\quad Conf(HI) := Support(HI)/Support(antecedent(HI))$
14 $\quad\lfloor\quad T_{HI} := T_{HI} \setminus t$;; Remove t from T_{HI}

3.6 Tabular Data

Tabular data in statistics correspond to aggregates of data with respect to a few variables. We illustrate tabular data with an example. This example is taken from [50] (see also [51]).

Example 3.12 Let us consider a file where records include profession, town and salaries.

From the information in the file we can build a table with aggregate values in which in one dimension we have the different professions and in the other the different towns. Cells in the table correspond to the frequency of each pair *(profession, town)*. Table 3.3 represents this situation. So, the cell (M_2, P_3) is the number of people with profession P_3 living in municipality M_2.

Table 3.4 is a similar table, but in this case we have that in each cell we have the total salary received by people with a certain job and living in a certain place.

In both tables we have totals for rows and columns, and a grand total for the full table.

The publication of data in aggregated form does not avoid disclosure problems. We describe below different types of attacks on the tables described in the previous example. Although in our attacks we use the two tables, it is not necessary that the two tables are published by an office. Public information can be sufficient to know the frequency in a particular cell. This is the case, for example, if there is a single

Table 3.3 Two-dimensional table of frequencies (adapted from [50])

	P_1	P_2	P_3	P_4	P_5	Total
M_1	2	15	30	20	10	77
M_2	72	20	1	30	10	133
M_3	38	38	15	40	5	136
Total	112	73	46	90	25	346

Table 3.4 Two-dimensional table of magnitudes (adapted from [50])

	P_1	P_2	P_3	P_4	P_5	Total
M_1	360	450	720	400	360	2290
M_2	1440	540	22	570	320	2892
M_3	722	1178	375	800	363	3438
Total	2522	2168	1117	1770	1043	8620

doctor in a town, or if cells contain information about companies classified by sectors and we know the companies in a certain sector located in a certain town.

- **External attack**. Combining the information of the two tables the adversary is able to infer some sensitive information. For example, combining the information in Tables 3.3 and 3.4 an adversary can find that the single person working as P_3 in town M_2 has a salary of 22.
- **Internal attack**. A person whose data is in the database is able to use the information of the tables to infer some sensitive information about other individuals. For example, a person whose work is P_1 and lives in M_1 can attack the data using his own salary. For example, if there are only two doctors in a town, each one will be able to find the salary of the other.
- **Internal attack with dominance**. This is an internal attack where a contribution of one person, say P_0, in a cell is so high that permits P_0 to obtain accurate bounds of the contribution of the others. For example, let us consider the 5 people with profession P_5 and living in M_3. If one of them has a salary of 350, then it is clear that the salary of the other four is at most 13.

Tabular data protection focus on the protection of tables so that these type of attacks cannot be done.

Protection in tabular data typically consists of removing or modifying some of the elements in the table in a way that the adversary cannot infer these values or can only infer them with sufficient uncertainty.

Methods for tabular data need first to determine which cells are sensitive. There are a few rules for this. They are known as sensitivity rules. We describe the most common ones below. These rules are applied to each cell independently. We will give formulations of these rules using $c_1, \ldots, c_t$ to denote the t contributions to the cell. We assume that all contributions are positive ($c_i > 0$).

- **Rule (n, k)-dominance**. A cell is sensitive when n contributions represent more than the k fraction of the total. That is, the cell is sensitive when

$$\frac{\sum_{i=1}^{n} c_{\sigma(i)}}{\sum_{i=1}^{t} c_i} > k$$

  where $\{\sigma(1), \ldots, \sigma(t)\}$ is a permutation of $\{1, \ldots, t\}$ such that $c_{\sigma(i-1)} \geq c_{\sigma(i)}$ for all $i = \{2, \ldots, t\}$ (i.e., $c_{\sigma(i)}$ is the ith largest element in the collection $c_1, \ldots, c_t$). This rule is used with $n = 1$ or $n = 2$ and $k > 0.6$.

- **Rule pq**. This rule is also known as the prior/posterior rule. It is based on two positive parameters p and q with $p < q$. Prior to the publication of the table, any intruder can estimate the contribution of contributors within the q percent. Then, a cell is considered sensitive if an intruder on the light of the released table can estimate the contribution of a contributor within p percent.

  Among all contributors, the best estimation can be done by the second largest contributor on the first largest contributor. Let us consider estimations of values for contributor r by the contributor s. Although derivation is general for any r and s ($r \neq s$), we will apply this to $s = \sigma(2)$ (to denote the second largest contributor) and $r = \sigma(1)$ (to denote the first larger contributor).

  Note that in general, due to the q percent, the estimation of the lower bound of any contributor i is lower than $c_i - (q/100)c_i$. I.e.,

$$lower\text{-}estimation(c_i) < c_i - \frac{q}{100}c_i.$$

  Then, the upper bound estimation of c_r computed by c_s is

$$upper\text{-}estimation(c_r) = \sum_{i=1}^{t} c_i - \sum_{i \neq s,r} lower\text{-}estimation(c_i) - c_s$$

$$= \sum_{i=1}^{t} c_i - \sum_{i \neq s,r} \left(c_i - \frac{q}{100}c_i\right) - c_s$$

$$= \sum_{i=1}^{t} c_i - \sum_{i \neq r} c_i + \frac{q}{100} \sum_{i \neq s,r} c_i$$

$$= c_r + \frac{q}{100} \sum_{i \neq s,r} c_i$$

  A cell will be sensitive if the estimations are inside the p percent interval. That is,

$$estimation(c_i) \in \left[c_i - \frac{p}{100}c_i, c_i + \frac{p}{100}c_i\right].$$

So, the upper estimation of c_r is sensitive if smaller than $c_r + \frac{p}{100}c_r$. Using the expression of the upper estimation of c_r we have that the cell is sensitive if

$$upper\text{-}estimation(c_r) = c_r + \frac{q}{100} \sum_{i \neq s,r} c_i < c_r + \frac{p}{100}c_r$$

So, equivalently, the cell is sensitive if

$$q \sum_{i \neq s,r} c_i < p \cdot c_r.$$

Taking into account what we have stated above that the better estimation is the one of the second largest contributor on the first largest contributor, we need to apply this expression with $s = \sigma(2)$ and $r = \sigma(1)$. This results into the following expression for the pq rule

$$q \sum_{i=3}^{t} c_{\sigma(i)} < pc_{\sigma(1)}. \tag{3.1}$$

That is, we check this condition for all cells and the cells that satisfy the inequality are sensitive.

We have developed this rule starting with an estimation of the lower bound and then applying the p percent on the upper bound. We can start with the upper bound and then apply the p percent on the lower bound. We would obtain the same Eq. 3.1.

- **Rule p%**. This rule can be seen as a special case of the previous rule when no prior knowledge is assumed on any cell. Because of that, it can be seen as equivalent to the previous rule with $q = 100$. Using Eq. 3.1 and using $q = 100$ we would say that a cell is sensitive when the contributions satisfy:

$$\sum_{i=3}^{t} c_{\sigma(i)} < \frac{p}{100} c_{\sigma(1)}. \tag{3.2}$$

According to [52], nowadays the pq rule and the p rule are preferred. Reference [53] discusses in details the drawbacks of the (n, k)-dominance rule. Another discussion on the rules for determining sensitive cells and some of its drawbacks can be found in [54]. Reference [54] proposes an alternative rule based on entropy. Reference [52] also discusses the rules and derive the expressions for the computation of the sensitivity of the rules. We have used them to derive the expressions here.

Table 3.5 Two-dimensional table of magnitudes with suppressed cells (primary and secondary suppressions) (adapted from [50] and [51])

	P_1	P_2	P_3	P_4	P_5	Total
M_1	360	450		400		2290
M_2	1440	540		570		2892
M_3	722	1178	375	800	363	3438
Total	2522	2168	1117	1770	1043	8620

Once sensitive cells are determined, we can apply a data protection method for tabular data. Following [51] we classify methods for tabular data protection into perturbative and non-perturbative. This classification appears again in Chap. 6 when we classify masking methods for standard databases.

- **Perturbative**. The values of the cells are modified and an alternative table where some cells contain inaccurate values is published. The two main methods are controlled rounding and controlled tabular adjustment. Rounding consists of replacing all values by multiples of a given base number. The controlled tabular adjustment consists of finding the nearest table (given an appropriate distance) of the table to be protected. We explain this approach below (Sect. 3.6.2).

  These approaches are known as post-tabular methods to distinguish them with pre-tabular perturbation. In such methods, noise is added before the table is prepared. See e.g. [55] for details.
- **Non-perturbative**. The table does not include any inaccurate value. Privacy is achieved by means of suppressions and modifications on the structure of the table. We explain below cell suppression for tables (Sect. 3.6.1).

3.6.1 Cell Suppression

Methods for cell suppression consist of removing the values in the sensitive cells. Nevertheless, this is not enough. Sensitive values should not be inferred from the published data and due to the fact that tables include marginals, it is not enough to remove the values that lead to disclosure.

Note that removing the value of the cell (M_2, P_3) in Table 3.4 does not avoid the disclosure. The adversary can use the subtotals to find the value of the cell. A way to ensure anonymity for cell (M_2, P_3) is to suppress the value of this cell and also the ones of other cells. In this case we have primary and complementary (or secondary) suppression. The suppression of the value in (M_2, P_3) is a primary suppression and the one of other cells secondary suppression. With secondary suppression, we can obtain something like Table 3.5. In this case cells (M_1, P_3), (M_1, P_5), and (M_2, P_5) are secondary suppressions.

This problem can be formulated in terms of an optimization problem which consists of minimizing the deletions (secondary cell suppression) in a way that the constraints are satisfied. The expression to be minimized is expressed in terms of an objective function (which permits different types of costs). The input elements of the problem are detailed below.

- **Positive table with n cells.** This is expressed by means of values $a_i \geq 0$ for $i = 1, \ldots, n$. In addition we have some linear relationships expressed in a matrix form $Aa = b$. Note that these relationships permit us to express marginals of the table (as well as other conditions, if required).
- **Weights of each cell.** It corresponds to the cost of modifying or removing each cell. These weights are denoted by w_i for $i = 1, \ldots, n$ and are such that $w \geq 0$. When all cells have equal weight $w_i = 1$ for all i.
- **Upper and lower bounds of a_i.** They are the bounds known by the intruders on the values of the cells (i.e., the previous knowledge). We denote them by kup_i and klo_i for $i = 1, \ldots, n$. Naturally, $klo \leq a \leq kup$. Default values are $(0, M)$ for positive tables (with M a large value) and $(-M, M)$ for arbitrary tables.
- **Set of sensitive cells $\mathcal{P}$.** They are a subset of all cells. That is, $\mathcal{P} \subseteq \{1, \ldots, n\}$.
- **Upper and lower protection levels.** For each cell p in $\mathcal{P}$ we denote its upper and lower protection level by up_p and lo_p. Then, we expect the estimated value of cell p, say x_p, to be outside the interval $[a_p - lo_p, a_p + up_p]$ (either above the upper limit or below the lower limit).

In order to formulate the problem, we need an additional set of variables.

- **Variables for suppression.** Variables y_i for all cells $i = 1, \ldots, n$ that establish whether a cell is suppressed or not. Formally, y_i take values in $\{0, 1\}$ and where $y_i = 1$ means that the ith cell has to be suppressed and $y_i = 0$ means that the cell is not modified. The goal of the problem is to determine the set of cells to be deleted, or in other words, to find the secondary suppressions $\mathcal{S}$ to be added to the primary suppressions $\mathcal{P}$. As all these cells will have $y_i = 1$, we have that

$$\mathcal{P} \cup \mathcal{S} = \{i \,|\, y_i = 1\}.$$

- **Variables for estimations.** Variables $x^{l,p}$ and $x^{u,p}$ for each primary cell p denote the lower and upper estimation of the intruders for primary cells $p \in \mathcal{P}$. So, $x^{l,p}$ will be the minimum of possible values x_p and $x^{u,p}$ will be the maximum of possible values x_p.

  The formulation of the problem is based on the fact that for each set of primary and secondary suppressions intruders can establish these lower and upper bounds for primary cells. These bounds $x^{l,p}$ and $x^{u,p}$ can be computed as follows:

- **Lower estimation**. For all $p \in \mathscr{P}$, the lower estimation is the solution of the following optimization problem

$$x^{l,p} = \min x_p$$
$$\text{subject to} \quad Ax = b$$
$$kloi \le x_i \le kup_i \text{ for all } i \in \mathscr{P} \cup \mathscr{S}$$
$$x_i = a_i \text{ for all } i \notin \mathscr{P} \cup \mathscr{S}$$

(3.3)

That is, the lower estimation is the lowest value that can be achieved that satisfies the linear relationships ($Ax = b$), the constraints on the previous knowledge of the intruder ($kloi \le x_i \le kup_i$), and the fact that undeleted cells keep their original value ($x_i = a_i$ if $i \notin \mathscr{P} \cup \mathscr{S}$).

- **Upper estimation**. For all $p \in \mathscr{P}$, the upper is found solving the following problem, which is similar to the one above for the lowest estimation

$$x^{u,p} = \max x_p$$
$$\text{subject to} \quad Ax = b$$
$$kloi \le x_i \le kup_i \text{ for all } i \in \mathscr{P} \cup \mathscr{S}$$
$$x_i = a_i \text{ for all } i \notin \mathscr{P} \cup \mathscr{S}$$

(3.4)

Let us now consider the formulation of the problem for cell suppression. This formulation uses $d^{l,p}$ and $d^{n,p}$ instead of $x^{l,p}$ and $x^{u,p}$. $d^{l,p}$ is the difference between the lower estimation and the real value of the cell $p \in \mathscr{P}$, and $d^{n,p}$ is the difference between the upper estimation and the real value of the cell p (i.e., a_p).

The goal of the protection is that after deletion $x^{l,p}$ and $x^{u,p}$ are out of the interval defined by the lower and upper protection levels. I.e., $x^{l,p} \le a_p - lo_p$ and $a_p + up_p \le x^{u,p}$ for all $p \in \mathscr{P}$.

Using $d^{l,p}$ and $d^{u,p}$ these inequalities are equivalently expressed as $d^{l,p} \le -lo_p$ and $up_p \le d^{u,p}$. These equations are included in the mathematical formulation below.

The mathematical problem also includes constraints so that the lower and upper constraints are within lower and upper limits of a_i known by intruders. I.e., if $y_i = 1$ then $kloi - a_i \le d^{l,i} \le kup_i - a_i$ or equivalently with $x^{l,i} = a_i + d^{l,i}$ this corresponds to $kloi \le x^{l,i} \le kup_i$, and similarly for the upper difference $d^{u,i}$ we require $kloi - a_i \le d^{u,p} \le kup_i - a_i$.

In addition, we need that estimates x^l and x^u satisfy the linear constraints. That is, $Ax^l = b$ and $Ax^u = b$. This constraint in terms of d^l and d^u corresponds to $Ad^l = 0$ and $Ad^u = 0$.

$$\min \sum_{i=1}^{n} w_i y_i$$

subject to

$$Ad^l = 0$$
$$(klo_i - a_i)y_i \leq d^{l,i} \leq (kup_i - a_i)y_i \quad \text{for all } i = 1, \ldots, n$$
$$d^{l,p} \leq -lo_p \quad \text{for all } p \in \mathscr{P}$$

$$Ad^u = 0$$
$$(klo_i - a_i)y_i \leq d^{u,i} \leq (kup_i - a_i)y_i \quad \text{for all } i = 1, \ldots, n$$
$$d^{u,p} \geq up_p \quad \text{for all } p \in \mathscr{P}$$

$$y_i \in \{0, 1\} \quad \text{for } i = 1, \ldots, n$$

Discussion on methods to solve this optimization problem can be found in e.g. [51].

3.6.2 Controlled Tabular Adjustment

This approach consists of replacing the original table by another one that is similar, and that satisfies a set of requirements. Here there will be no suppressed cells but cells with noisy values.

This approach was independently developed by Dandekar and Cox [56] and Castro [57], and defined in extended form in [58]. In [57], the approach was named minimum-distance controlled perturbation method.

The formulation of the solution requires a set of variables. They are the following ones. Note that these variables are exactly the ones used in Sect. 3.6.1.

- **Positive table with n cells.** We use $a_i \geq 0$ with $i = 1, \ldots, n$ to denote the values of the cells. In addition we have some linear relationships expressed in a matrix form $Aa = b$. Note that these relationships permit us to express marginals of the table as well as other relationships.
- **Weights of each cell.** The cost of modifying or removing each cell is denoted by w_i for $i = 1, \ldots, n$ and are such that $w \geq 0$.
- **Upper and lower bounds of a_i.** These bounds correspond to what is known by the intruder. They are denoted, respectively, by kup_i and klo_i for $i = 1, \ldots, n$. Naturally, the bounds should be such that $kup \leq a \leq klo$. Default values are $(0, M)$ for positive tables (and where M is a large value) and $(-M, M)$ for arbitrary tables.
- **Set of sensitive cells** $\mathscr{P}$. They are the primary cells and are a subset of all cells. That is, $\mathscr{P} \subseteq \{1, \ldots, n\}$.

- **Upper and lower protection levels**. For each primary cell p (i.e., p in $\mathscr{P}$) we have the bounds of the protection. These bounds are, respectively, up_p and lo_p for $p \in \mathscr{P}$. Then, we expect the protected value of cell p, that is x_p, to be outside the interval $[a_p - lo_p, a_p + up_p]$ (either above the upper limit or below the lower limit).

The goal is to find a table near to a where near is defined in terms of a distance. The new table is denoted by x which corresponds to the values $x_1, \ldots, x_n$ (one value for each cell). The distance between x and a is expressed as $||x - a||_{L_s(w)}$ and corresponds to the following expression

$$||x - a||_{L_s(w)} = \sum_i w_i |x^i - a^i|^s$$

for x^i and a^i the components of x and a, respectively.

Note that in this formulation x_p is the value of cell p in the new matrix, and this is different to the role of x_p in Sect. 3.6.1, where x_p were estimations for the cells p in $\mathscr{P}$ of the suppressed values.

Controlled tabular adjustment is formulated in terms of the following optimization problem. We can see that in this problem we minimize the distance between x and a subject to the fact that the linear relationship on A (as e.g. the marginals) are not modified ($Ax = b$) and that the values of the new matrix are within the bounds known by the intruder ($klo \le x \le kup$) but outside the sensitive intervals (either $x_p \le ap - lo_p$ or $x_p \ge a_p + up_p$).

$$
\begin{aligned}
&\min_x \ ||x - a||_{L_s(w)} \\
&s.t. \ \ Ax = b \\
&\quad\quad klo \le x \le kup \\
&\quad\quad x_p \le a_p - lo_p \text{ or } x_p \ge a_p + up_p \text{ for } p \in \mathscr{P}
\end{aligned}
$$

This more general definition allows x_p to be either below the limits of the interval or over the limits of the interval. Nevertheless, this formulation results into a difficult optimization problem. An alternative is to allow values only on one side of the interval. In this case the problem is a continuous optimization problem that is easier to solve. See [51,58,59] for details.

As stated above, this approach was introduced independently by Dandekar, Cox and Castro. Dandekar and Cox [56] used L_1 distance and Castro [57,60] L_2. A unified definition was given by Castro in [58]. This latter work also studies the disclosure risk of this approach. Reference [59] extended this study by means of an exhaustive empirical analysis for different tables using both L_1 and L_2 distances. Reference [59] summarizes the study on the risk stating that "if the attacker does not have good knowledge on the original data, he/she could hardly obtain good estimates of the sensitive cells, in general. However, if the attacker has good information about the protection levels and which are the sensitive cells, or he/she knows the lower and upper bounds of the optimization problem (which is a stronger condition), then the method has a high disclosure risk". An analysis of the risk and data utility for these methods seems to indicate that L_2 is more suitable than L_1 (see [59,61]).

References

1. Riera, C.: Te deix, amor, la mar com a penyora, Edicions 62
2. Domingo-Ferrer, J.: A three-dimensional conceptual framework for database privacy. In: Proceedings of SDM 2007. LNCS, vol. 4721, pp. 193–202 (2007)
3. Verykios, V.S., Damiani, M.L., Gkoulalas-Divanis, A.: Privacy and security in spatiotemporal data and trajectories. In: Giannotti, F., Pedreschi, D. (eds.) Mobility, Data Mining and Privacy: Geographic Knowledge Discovery, pp. 213–240. Springer, Heidelberg (2008)
4. Bonchi, F., Saygin, Y., Verykios, V.S., Atzori, M., Gkoulalas-Divanis, A., Kaya, S.V., Savaş, E.: Privacy in spatiotemporal data mining. In: Giannotti, F., Pedreschi, D. (eds.) Mobility, Data Mining and Privacy: Geographic Knowledge Discovery, pp. 297–333. Springer, Heidelberg (2008)
5. Fung, B.C.M., Wang, K., Fu, A.W.-C., Yu, P.S.: Introduction to Privacy-Preserving Data Publishing: Concepts and Techniques. CRC Press (2011)
6. http://www.dssresources.com/newsletters/66.php. Accessed Jan 2017
7. http://www.census.gov. Accessed Jan 2017
8. Bache, K., Lichman, M.: UCI Machine Learning Repository. School of Information and Computer Science, University of California, Irvine (2013)
9. Willenborg, L., de Waal, T.: Elements of Statistical Disclosure Control. Lecture Notes in Statistics. Springer, Heidelberg (2001)
10. Atallah, M., Bertino, E., Elmagarmid, A., Ibrahim, M., Verykios, V.: Disclosure limitation of sensitive rules. In: Proceedings of IEEE Knowledge and Data Engineering Exchange Workshop (KDEX) (1999)
11. Atzori, M., Bonchi, F., Giannotti, F., Pedreschi, D.: Anonymity preserving pattern discovery. VLDB J. **17**, 703–727 (2008)
12. Liu, K., Kargupta, H., Ryan, J.: Random projection based multiplicative data perturbation for privacy preserving data mining. IEEE Trans. Knowl. Data Eng. **18**(1), 92–106 (2006)
13. Uribe, K.: Bitartean heldu eskutik. Susa (2001)
14. Torra, V.: Towards knowledge intensive data privacy. In: Proceedings of DPM 2010. LNCS, vol. 6514, pp. 1–7 (2011)
15. Clifton, C., Tassa, T.: On syntactic anonymity and differential privacy. In: Proceedings of ICDE Workshops (2013)
16. Li, N., Qardaji, W., Su, D.: On sampling, anonymization, and differential privacy: or, k-anonymization meets differential privacy. In: Proceedings of 7th ASIACCS 2012 (2011)
17. Dwork, C.: Differential privacy. In: Proceedings of ICALP 2006. LNCS, vol. 4052, pp. 1–12 (2006)
18. Yi, X., Paulet, R., Bertino, E.: Homomorphic Encryption and Applications. Springer, Heidelberg (2014)
19. Evfimievski, A., Fagin, R., Woodruff, D.: Epistemic privacy. J. ACM **58**, 1 (2010)
20. Deutsch, A., Papakonstantinou, Y.: Privacy in database publishing. In: Proceedings of ICDT 2005. LNCS, vol. 3363, pp. 230–245 (2005)

21. Jajodia, S., Meadows, C.: Inference problems in multilevel secure database management systems. In: Abrams, M.D., Jajodia, S., Podell, H.J. (eds.) Information Security, pp. 570–584. IEEE (1995)
22. Dwork, C., Nissim, N.: Privacy-preserving data mining on vertically partitioned databases. In: Proceedings of CRYPTO 2004. LNCS, vol. 3152, pp. 528–544 (2004)
23. Blum, A., Dwork, C., McSherry, F., Nissim, K.: Practical privacy: the SuLQ framework. In: Proceedings of PODS 2005, pp. 128–138 (2005)
24. Dwork, C.: Differential privacy: a survey of results. In: Proceedings of TAMC 2008. LNCS, vol. 4978, pp. 1–19 (2008)
25. Dwork, C.: The differential privacy frontier. In: Proceedings of TCC 2009. LNCS, vol. 5444, pp. 496–502 (2009)
26. Wasserman, L., Zhou, S.: A statistical framework for differential privacy. J. Am. Stat. Assoc. **105**(489), 375–389 (2010)
27. Sarathy, R., Muralidhar, K.: Some additional insights on applying differential privacy for numeric data. In: Proceedings of PSD 2010. LNCS, vol. 6344, pp. 210–219 (2010)
28. Bambauer, J., Muralidhar, K., Sarathy, R.: Fool's gold: an illustrated critique of differential privacy. Vanderbilt J. Entertainment Technol. Law (2014, in press)
29. Xiao, X., Wang, G., Gehrke, J.: Differential privacy via wavelet transforms. IEEE Trans. Knowl. Data Eng. **23**(8), 1200–1214 (2009)
30. Soria-Comas, J., Domingo-Ferrer, J.: Sensitivity-independent differential privacy via prior knowledge refinement. Int. J. Uncertainty Fuzziness Knowl. Based Syst. **20**(6), 855–876 (2012)
31. Korolova, A., Kenthapadi, K., Mishra, N., Ntoulas, A.: Releasing search queries and clicks privately. In: Proceedings of WWW (2009)
32. Task, C., Clifton, C.: A guide to differential privacy theory in social network analysis. In: Proceedings of 2012 IEEE/ACM International Conference on Advances in Social Networks Analysis and Mining (2012)
33. Johansson, F.D., Frost, O., Retzner, C., Dubhashi, D.: Classifying large graphs with differential privacy. In: Proceedings of MDAI 2015. LNCS, vol. 9321, pp. 3–17 (2015)
34. Yao, A.C.: Protocols for secure computations. In: Proceedings of 23rd IEEE Symposium on Foundations of Computer Science, Chicago, Illinois, pp. 160–164 (1982)
35. Lindell, Y., Pinkas, B.: Privacy preserving data mining. In: Crypto 2000. LNCS, vol. 1880, pp. 20–24 (2000)
36. Lindell, Y., Pinkas, B.: Privacy preserving data mining. J. Cryptol. **15**, 3 (2002)
37. Bunn, P., Ostrovsky, R.: Secure two-party k-means clustering. In: Proceedings of CCS 2007, pp. 486–497. ACM Press (2007)
38. Vaidya, J., Clifton, C.W., Zhu, Y.M.: Privacy Preserving Data Mining. Springer, Heidelberg (2006)
39. Kantarcioglu, M.: A survey of privacy-preserving methods across horizontally partitioned data. In: Aggarwal, C.C., Yu, P.S. (eds.) Privacy-Preserving Data Mining: Models and Algorithms, pp. 313–335. Springer, Heidelberg (2008)
40. Hajian, S.: Simultaneous discrimination prevention and privacy protection in data publishing and mining, Ph.D. Dissertation, Universitat Rovira i Virgili (2013)
41. Hajian, S., Domingo-Ferrer, J., Martínez-Ballesté, A.: Rule protection for indirect discrimination prevention in data mining. In: Proceedings of MDAI 2011. LNCS, vol. 6820, pp. 211–222 (2011)
42. Bertino, E., Lin, D., Jiang, W.: A survey of quantification of privacy preserving data mining algorithms. In: Aggarwal, C.C., Yu, P.S. (eds.) Privacy-Preserving Data Mining: Models and Algorithms, pp. 183–205. Springer, Heidelberg (2008)
43. Haritsa, J.R.: Mining association rules under privacy constraints. In: Aggarwal, C.C., Yu, P.S. (eds.) Privacy-Preserving Data Mining: Models and Algorithms, pp. 239–266. Springer, Heidelberg (2008)

44. Abul, O.: Knowledge hiding in emerging application domains. In: Bonchi, F., Ferrari, E. (eds.) Privacy-Aware Knowledge Discovery, pp. 59–87. CRC Press (2011)
45. Dasseni, E., Verykios, V.S., Elmagarmid, A.K., Bertino, E.: Hiding association rules by using confidence and support (2001)
46. Verykios, V.S., Elmagarmid, A.K., Bertino, E., Saygın, Y., Dasseni, E.: Association rule hiding. IEEE Trans. Knowl. Data Eng. 16, 434–447 (2004)
47. HajYasien, A.: Preserving privacy in association rule hiding, Ph.D. Dissertation, Griffith University (2007)
48. Verykios, V.S.: Association rule hiding methods. WIREs Data Min. Knowl. Discov. 3, 28–36 (2013). doi:10.1002/widm.1082
49. Verykios, V.S., Elmagarmid, A.K., Bertino, E., Saygın, Y., Dasseni, E.: Association rule hiding, version Jan. 7, 2003, of [45] (2003)
50. Castro, J.: Taules estadístiques i privadesa. Universitat Estiu URV, La lluita contra el big brother (2007)
51. Castro, J.: Recent advances in optimization techniques for statistical tabular data protection. Eur. J. Oper. Res. 216, 257–269 (2012)
52. Daalmans, J., de Waal, T.: An improved formulation of the disclosure auditing problem for secondary cell suppression. Trans. Data Priv. 3, 217–251 (2010)
53. Robertson, D.A., Ethier, R.: Cell Suppression: Experience and Theory. LNCS, vol. 2316, pp. 8–20 (2002)
54. Domingo-Ferrer, J., Torra, V.: A critique of the sensitivity rules usually employed for statistical table protection. Int. J. Uncertainty Fuzziness Knowl. Based Syst. 10(5), 545–556 (2002)
55. Evans, T., Zayatz, L., Slanta, J.: Using noise for disclosure limitation of establishment tabular data. J. Official Stat. 14(4), 537–551 (1998)
56. Dandekar, R.A., Cox, L.H.: Synthetic tabular data: an alternative to complementary cell suppression, manuscript. Energy Information Administration, US Department of Energy (2002)
57. Castro, J.: Computational experiments with minimum-distance controlled perturbation methods. In: Proceedings of PSD 2004. LNCS, vol. 3050, pp. 73–86 (2004)
58. Castro, J.: Minimum-distance controlled perturbation methods for large-scale tabular data protection. Eur. J. Oper. Res. 171, 39–52 (2006)
59. Castro, J.: On assessing the disclosure risk of controlled adjustment methods for statistical tabular data. Int. J. Uncertainty Fuzziness Knowl. Based Syst. 20, 921–942 (2012)
60. Castro, J.: Internal communication to partners of the European Union IST-2000-25069 CASC project (2002)
61. Castro, J.: Comparing L1 and L2 distances for CTA. In: Proceedings of PSD 2012. LNCS, vol. 7556, pp. 35–46 (2012)

User's Privacy

4

Jo vinc d'un silenci
antic i molt llarg,
de gent sense místics
ni grans capitans,
que viuen i moren
en l'anonimat.

Raimon, Jo vinc d'un silenci, 1975.

According to our definition in Sect. 3.1.1, we have user's privacy when users have an active role in ensuring their own privacy. This is the case of Claudia and Berta in Examples 3.3 and 3.5. In this chapter we review a few tools that help users with this purpose.

Methods for user's privacy can be classified in two main classes according to their objective. We have

- Methods that protect the identity of the user, and
- Methods that protect the data of the user.

This distinction is clear in the case of Claudia (Example 3.3) sending queries to the search engine. The first objective is fulfilled when the search engine receives Claudia's query but is not able to link this query to Claudia. The second objective is fulfilled if Claudia is able to get the information she needs from the search engine without disclosing her interests to the search engine.

In contrast, in Example 3.5, Berta mainly focuses on protecting her identity. The ultimate goal is that the government is unable to link the message to her. Nothing is required about the privacy of the content of the message.

© Springer International Publishing AG 2017
V. Torra, *Data Privacy: Foundations, New Developments and the Big Data Challenge*, Studies in Big Data 28,
DOI 10.1007/978-3-319-57358-8_4

We will consider methods for user privacy in two contexts that roughly correspond to the two cases described above.

First, we will consider user privacy in communications. In this case, when a user A sends a message m to B, we may want to hide who is the sender (or the recipient) of the message, and to hide the content of the message.

Secondly, we will consider user privacy in information retrieval. In this case, we may want to hide who is sending the query or hide the query itself.

The chapter finishes with a brief section considering other contexts.

4.1   User Privacy in Communications

This section on tools for user privacy in communications is divided in two parts. First, we focus on the tools to avoid the disclosure of the identity of a user. They are tools for user anonymity. Then, we focus on tools to protect the data of the user and, more specifically, to ensure unobservability.

For a survey on this topic we refer the reader to [1]. The main systems for privacy in communications can also be found in Section 8 of [2].

4.1.1   Protecting the Identity of the User

Approaches for anonymous communications can be classified in two main classes as follows.

- **High-latency anonymity systems**. They correspond to applications in which interaction is not needed. This is the case of email. Anonymous systems of this class include mix networks
- **Low-latency anonymity systems**. In this case interaction is needed, and we need response in real-time. Web browsing is the most typical application. Anonymous systems of this class include onion routing and crowds

We present these systems below.

4.1.1.1   High-Latency Anonymity Systems

Mix networks were introduced by Chaum (1981) in [3] to unlink the sender and the recipient of a message. They are an example of high-latency anonymity systems. A proxy server (a *mix*) receives an encrypted message from the sender and forwards it to the recipient. Sender and recipient are unlinked when the server receives and sends messages from different senders to different recipients after shuffling them.

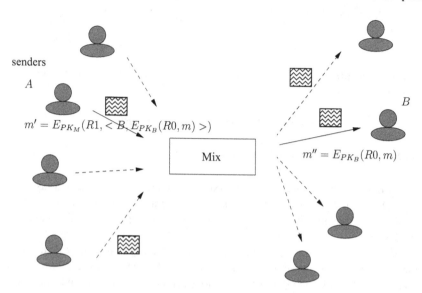

Fig. 4.1 Mixes: a message m is sent to the mix as $m' = E_{PK_M}(R1, < B, E_{PK_B}(R0, m) >)$ and from the mix to the addressee as $m'' = E_{PK_B}(R0, m)$

The network works using a protocol based on public-key cryptography.[1] We will use $E_K(m)$ to denote the encryption using the key K of the message m, and D_K to denote the decryption using key K. The whole process of sending a message using a mix is described below. Figure 4.1 illustrates this process. Each user B has a public key PK_B and a private key SK_B, and the mix has a public key PK_M and a private key SK_M.

- **Step 1. Message preparation**. A sender wants to send a message m to recipient B. Using the public key PK_M of the mix and the public key PK_B of B computes:

$$m' = E_{PK_M}(R1, < B, E_{PK_B}(R0, m) >)$$

where $R0$ and $R1$ are random strings, B denotes the address of the recipient, (a, b) denotes the message with the string a concatenated with b, and $< a, b >$ denotes a message with the pair a and b.
- **Step 2. Message sent**. The message m' is sent to the mix. The mix decrypts m' using its private key.

$$m^* = D_{SK_M}(m') = D_{SK_M}(E_{PK_M}(R1, < B, E_{PK_B}(R0, m) >)) \quad (4.1)$$
$$= (R1, < B, E_{PK_B}(R0, m) >) \quad (4.2)$$

---

[1]Cryptographic algorithm which requires two separate keys one of which is private and one of which is public. Also known as asymmetric cryptography.

Then, the mix discards the random string $R1$, and sends the message $E_{PK_B}(R0, m)$ to B. Let m'' denote this message.

$$m'' = E_{PK_B}(R0, m)$$

- **Step 3. Reception**. The recipient B uses his private key to decrypt the message. That is, it computes:

$$D_{SK_B}(m'') = D_{SK_B}(E_{PK_B}(R0, m)) = (R0, m).$$

Then, it discards the random string $R0$ to find m.

In this description we have considered a single mix, but a series of them (a cascade) can also be used. The advantage of a cascade is that all mixes have to collaborate to break the anonymity. In other words, a single non-compromised mix is able to provide secrecy.

We do not give here any explanation on when the messages are sent from the mix to the recipient or, in general, to other mixes. How and when messages are fired by a mix is defined by flushing algorithms. Methods can flush every t_s seconds, or when t_m messages are accumulated in the mix. See e.g. [4] for a discussion. Note that if the mix applies a first-in first-out approach for sending the messages it may be possible for an intruder to link senders and recipients.

There have been other systems as anon.penet.fi (in Finland, from 1993–1996, a centralized remailer system), and Cypherpunk and Mixminion [5] remailers (distributed remailer systems based on mix networks). Mixminion is the most sophisticated remailer among these ones permitting e.g. recipients to reply the sender without including sender's address in the body of the message.

4.1.1.2 Low-Latency Anonymity Systems

Low-latency anonymity systems have been developed for real-time applications when it is not appropriate to have delays in the reception of the request by the recipient. Recall that flushing algorithms in mixes delay transmissions. In this section we review crowds and onion routing, two of the existing low-latency systems.

Crowds. Crowds were introduced in [6] to achieve anonymity of web transactions. A crowd is defined as a collection of users. Then, when a user needs to transmit a transaction, it is either submitted directly or passed to another member of the crowd. Anonymity comes from the fact that users send some of their transactions but also end transactions from other members of the crowd.

The process is described below. A graphical representation of the path followed by a request through a crowd is given in Fig. 4.2.

- **Step 1. Start**. The user starts the local process (this process is known as jondo), that will represent him in the crowd.

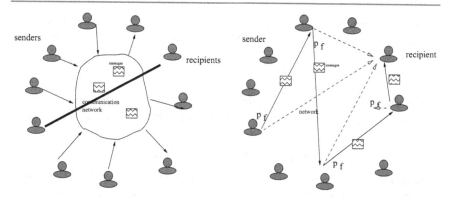

Fig. 4.2 Graphical representation of a crowd system. In the *left* the standard case of sending the request directly to the recipient. In the *right*, the request is processed by the crowd. Each process (jondo) forwards the request to another jondo with probability p_f and to the final recipient with probability $1 - p_f$. *Solid lines* represent actually used communication channels, *dotted lines* not used communication channels

- **Step 2. Contacting the server**. The jondo contacts a server called the blender to request admitance in the crowd.
- **Step 3. Admission**. The blender admits the jondo and sends him the information needed to participate in the crowd.
- **Step 4. Jondo as proxy**. The user selects this jondo just created as the web proxy for all services.
- **Step 5. Request sent**. Any request from the browser, is sent to the jondo.
- **Step 6. Processing**. When a jondo receives a request, it submits the request to the end server (the final destination of the request) with probability $1 - p_f$. Otherwise, with probability p_f, it forwards the request to another jondo. In this latter case, the receiver jondo is selected at random among possible ones. Note that this step is applied either for requests sent to the jondo by the user, or sent to the jondo by other jondos.

Onion routing. This is an approach for real-time and application independent anonymous communication. Both data and communication is made anonymous. It was defined in [7]. Tor [8,9] is a current implementation of onion routing, although its actual mechanism is more sophisticated than the one described here (using e.g. Diffie-Hellman key exchange).

The communication process follows the steps described below.

- **Step 1. Retrieval of known routers**. The user accesses a directory server that provides known routers and their current state.
- **Step 2. Construction of the path**. The list of routers is used to define a path. That is, an ordered list of at least three routers that will be traversed by the message.

- **Step 3. Onion.** A data structure, called onion, is built for the message. Each layer of the onion defines one hop in the route. Public-key cryptography is used for this purpose.

  Let $E_{PK}(m)$ denote the encryption using the public key PK of the message m. Let $< i, m >$ denote that we send message m to router i. Then, if the message m is sent through the path traversing nodes 3, 5, and 6 with public keys PK_3, PK_5, and PK_6, respectively, the onion for this message denoted by o will be something like:

  $$o = E_{PK_3}(< 5, E_{PK_5}(< 6, E_{PK_6}(m) >) >).$$

- **Step 4. Message passing.** The message is passed to the entry funnel, an onion router with a longstanding connection. The router peels off its layer, identifies the next hop, and sends the embedded onion to this onion router.

  Peeling off the layer consists of applying D_{SK}, the decryption mechanism, to the message. Here SK is the private key of the router.

  In the case above, router 3 is the entry funnel. This router applies D_{SK_3} to the onion o obtaining:

  $$D_{SK_3}(o) = D_{SK_3}(E_{PK_3}(< 5, E_{PK_5}(< 6, E_{PK_6}(m) >) >))$$
  $$= < 5, E_{PK_5}(< 6, E_{PK_6}(m) >) > .$$

  Then, the onion $o' = E_{PK_5}(< 6, E_{PK_6}(m) >)$ is passed to router 5.

- **Step 5. Message forwarded.** The same approach is applied subsequently by the other routers in the path.

  In this example, router 5 will decrypt the message and forward

  $$o'' =< 6, E_{PK_6}(m) >$$

  to router 6.

- **Step 6. Message delivery.** Once the message arrives to the last router in the path, this router (known as the exit funnel) using the information in m delivers the message to the appropriate address.

- **Step 7. On the reply.** The process is reversed for data moving back to the original sender. In this case, each router encrypts the data using the private keys. The recipient (the individual that initiated the communication) will decrypt the message using the original path and the public keys of the routers in the path.

  Following with the example above, the following answer will be received by the original sender:

  $$m' = E_{SK_3}(E_{SK_5}(E_{SK_6}(answer))).$$

  Applying the decryption mechanism to the message received in the appropriate order, we retrieve the answer:

  $$answer = D_{PK_6}(D_{PK_5}(D_{PK_3}(E_{SK_3}(E_{SK_5}(E_{SK_6}(answer)))))).$$

There are some similarities between onion routing and mixes, and more especifically with a cascade of mixes. Reference [7] discusses that one of the differences between onion routing and mixes is that the routers of the former "are more limited in the extent to which they delay traffic at each node". Another difference is that in onion routing, all routers are entry points, and traffic entering or exiting the nodes may not be visible.

4.1.2 Protecting the Data of the User

navia aut caput

If the only goal is to provide protection to the data, we can use cryptographic tools. A sender can use the public key of the recipient so that only the recipient can decrypt it with the corresponding private key.

In the remaining part of the section we present the solution of a problem that provides protection but also unobservability. Note that an alternative approach for achieving unobservability is dummy traffic. When communication is encrypted users can send fake data (but indistinguishable from real encrypted data) at appropriate times to confuse intruders and avoid traffic analysis (e.g. to avoid the disclosure of when and to whom data is sent).

The problem we consider is the one of the dinning cryptographers. The method also provides sender anonymity (and as such would fall in Sect. 4.1.1). Equivalently, it is also a secure multiparty computation of the function OR. This problem was first solved in [10]. Let us first review the original formulation of this problem.

Problem 4.1 [10] Three cryptographers are sitting down to dinner at their favorite three-star restaurant. Their waiter informs them that arrangements have been made with the maître d'hôtel for the bill to be paid anonymously. One of the cryptographers might be paying the dinner, or it might have been NSA (U.S. National Security Agency). The three cryptographers respect each other's right to make an anonymous payment, but they wonder if NSA is paying.

So, the problem is to know whether one of the cryptographers pay.

To solve this problem, the following protocol can be used. Figure 4.3 illustrates the steps in the case of one cryptographer paying (left) and in the case of none of them paying (right).

- **Step 1**. Each cryptographer flips a coin and shares its outcome with the crytographer on the right. Let us represent tails and heads by 1 and 0, respectively. Let $coin_i$ be the outcome of the coin of the ith cryptographer.

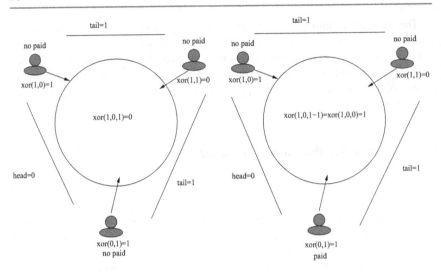

Fig. 4.3 Dining cryptographers: computation of the output when a cryptographer pays (*left*) and when none of them pay (*right*)

- **Step 2**. All cryptographers find whether the two coins they know about (the ones they flipped and the one their left-hand neighbor flipped) fell on the same side or not. Let us consider the ith cryptographer, then we can use the *xor* on the results of the two coins to represent cryptographer's computation:

$$c_i = xor(coin_i, coin_{(i-1)mod3}).$$

- **Step 3**. If a cryptographer is the payer, then the answer is the opposite of what is observed. Otherwise, the answer is what is observed. Formally, let us represent the answer of the ith cryptographer by c_i', then

$$c_i' = \begin{cases} c_i & \text{if the ith cryptographer did not pay the meal} \\ 1 - c_i & \text{if the ith cryptographer paid the meal.} \end{cases} \quad (4.3)$$

- **Step 4**. Then, let s be the sum of the values c_i'. If the sum is even, no one paid. If odd, one crytographer paid. The *xor* function can be used for this purpose. That is, the output is:

$$xor(c_1', c_2', c_3').$$

This protocol can be generalized to an arbitrary number of cryptographers. In that case, each cryptographer needs a secret bit with each other participant. Each cryptographer computes the sum modulo two (or the *xor* function of all the bits). Then, the ith cryptographer applies the function above to determine c_i' from c_i (Expression 4.3). Finally, as in Step 4 above, let s be the sum of the values c_i'. If the sum is even, no one paid. If odd, one cryptographer paid.

The main problems of this protocol are three: (i) malicious participants make the output useless; and (ii) for n participants we need n^2 communications (one for each pair of participants).

In addition, only one participant can transmit a bit at a time. Two bits from different participants would cancel each other and would not be detected.

4.2 User Privacy in Information Retrieval

Information retrieval is about querying a database to get some information. In this setting we can also consider the two scenarios seen in the previous section. In the first, the user wants to hide her identity and in the second the user wants to hide her query to the database. These two scenarios are described below.

4.2.1 Protecting the Identity of the User

Anonymous database search, also known as user-private information retrieval [11], is the area that studies how to ensure that the database receives the query without knowing who is the sender of the query. This problem can be addressed using techniques similar to onion routing and crowds (discussed in Sect. 4.1.1.2).

That is, a user posts queries in the peer-to-peer (P2P) community, and after a sequence of forwards, another user forwards the query to the database. When this other user receives the answer to the query, it forwards it to the interested user.

Queries posted in the P2P community are encrypted so that only members can read them. Different communication spaces can be used. Each communication space is a set of users and the cryptographic key they share. At any point of time, we have in the communication space a set of messages waiting to be processed. The use of several communication spaces decrease the risk of compromising keys and at the same time it implies that not all users can see the queries in plain text. Combinatorial configurations [11–13] can be used to find the optimal way to organize users and communication spaces. A survey on the use of combinatorial configurations for anonymous database search can be found in [14].

Algorithm 11 describes the procedure that users follow according to [15]. At each step the user selects a communication space and decrypts its content obtaining a queue of messages $M = (M_i)$. Then, each message M_i is processed, and a new set of messages are encrypted and posted to the communication space.

This algorithm should be repeated with frequency f. This frequency does not need to be the same for all users, but it should be higher or equal to the highest query submission frequency among the users. When all communication spaces have the same number of users (k) and all users are in the same number of communication spaces (r), the probability x of self-submission should be fixed to

$$x = 1/(r(k-1)+1).$$

Algorithm 11: Anonymous database search: P2P UPIR 2 protocol in [15] following [1].

Data: p: user
Result: Void

1 $l :=$ Select uniformly at random a communication space l of which p is member ;
2 $M :=$ Decrypt the content on l using the corresponding cryptographic key ;
3 **foreach** $M_i \in M$ **do**
4 | **if** M_i *is a query addressed to* p **then**
5 | | M_i is removed from the queue and forwarded to the server. The answer A from the server is encrypted and added at the end of the queue ;
6 | **else**
7 | | **if** M_i *is an answer to a query belonging to* p **then**
8 | | | M_i is read and removed from the queue
9 | | **else**
10 | | | M_i is left on the queue without action

11 **if** p *has a query* Q **then**
12 | With probability x, we define $p' = p$; and with probability $1 - x$ the user selects uniformly at random another user $p' \neq p$ on l ;
13 | p addresses Q to p' and writes Q to the end of the queue ;

4.2.2 Protecting the Query of the User

The most secure approaches (in terms of the privacy level achieved) to protect the queries of a user are studied in the area of Private Information Retrieval (PIR). In PIR, the database or search engine knows who the user is but nothing about the queries of the user. PIR methods, which are based on cryptographic protocols, are explained in Sect. 4.3. These methods require the database to collaborate with the users and run appropriate protocols.

There is a family of methods that do not require the collaboration of the database. Some are based on adding noise to the queries and others on exploiting the fact that not all queries are sensitive. We review these approaches below.

4.2.2.1 Noise Addition for Queries

TrackMeNot [16] and GooPIR [17] are implemented as agents in a browser to avoid that a search engine knows with certainty about the interests of a user. In addition to the queries sent by the user, the agents (two plug-ins) send fake queries. The rationale is that the search engine cannot distinguish between the real queries and the fake ones. In this way, any profile the search engine builds from the user will be noisy.

TrackMeNot [16] harvests query-like phrases from the web and sends them to the search engine. Queries are constructed using a set of RSS feeds from popular web sites as e.g. the New York Times, CNN, and Slashdot and a list of popular queries gathered from publicly available lists. The reply of the search engine is used for *refining* the query and submitting a new one. The approach implemented causes that each system evolves differently, due to different selections of terms from RSS feeds, terms selected from the search engine response, and the information returned by the search engine (that also evolves with time). In order to mimic user's behavior some links are clicked (selection is done in such a way to avoid revenues to the search engine) and queries are sent in batch similar to what users do.

GooPIR [17] follows a different strategy. While TrackMeNot builds fake queries, GooPIR adds fake terms to the real user queries. It uses a thesaurus to select a number of keywords with frequencies similar to the ones on the query, and then adds the selected keywords with OR to the real query. Then, the query is submitted to the search engine, and, finally, the results given by the search engine are filtered.

The effectiveness of such systems in ensuring privacy was considered in [18]. This work considers an adversarial search engine that uses users' search histories to determine which queries are generated by TrackMeNot and which are the real queries. Classification is based on machine learning techniques. The experiment was performed using 60 users from the AOL search logs. The authors state (see page 3 in [18]) that they achieved a 48.88% average accuracy for identifying user queries (predicted as user queries over real user queries), and a 0.02% of incorrectly classified fake queries (incorrectly predicted as belonging to TrackMeNot over the real TrackMeNot queries). The difference of identification between users was large: the range was between 10 and 100%.

4.2.2.2   DisPA: The Dissociating Privacy Agent

DisPA [19,20], which stands for Dissociating Privacy Agent, is an agent implemented as a proxy between the user and the search engine. The goal is to protect the queries of the user, but at the same time permit certain level of profiling by the search engine. In this way, there are privacy guarantees but also personalization on the results delivered by the search engine.

From the point of view of privacy, DisPA is based on the idea that what makes people unique are their set of interests (recall the discussion in Sect. 1.4 that individuals are multifaceted). That is, every person has a set of interests that considered in isolation are usually not enough to identify this person but all together permit the reidentification. In other words, for each interest, the anonymity set is usually large. It is the intersection of these interests what makes the person unique and makes the anonymity set small, if not a singleton.

Similarly, queries considered in isolation do not cause reidentification, but the intersection of them is what can make them linkable to the individual.[2] Based on these principles, DisPA *dissociates* the user identity into different virtual identities, each one with a particular interest. Then, each query is classified into one of these interests, and assigned to the corresponding virtual identity. Then, the query is submitted to

---

[2]Note the parallelism with social spheres [21,22] and the privacy problems in online social networks when these spheres are put in contact.

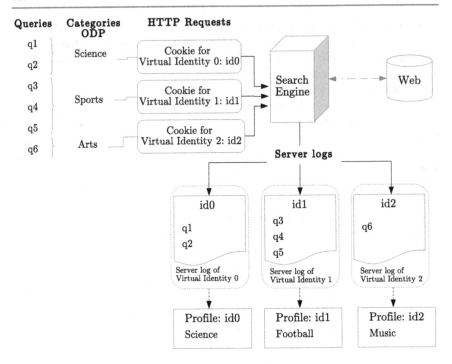

Fig. 4.4 DisPA architecture from [23]

the agent by this virtual identity, and the result of the query is received by this virtual identity, and then forwarded to the user. Figure 4.4 represents this process.

For example, if a user is interested in football and piano music. The agent will dissociate these queries and assign one virtual identity to football and another one to piano. Then, each query related to football will be submitted to the database by the corresponding virtual identity. Similarly, the other identity will be used to submit the other queries related to piano. In order to classify queries to the appropriate virtual identity, DisPA implements a classifier that uses the taxonomy in the Open Directory Project [24].

It is well known that the use of proper names in queries helps much in any attack. Because of that the DisPA agent implements a filter to detect entities like locations and personal names, unique identifiers and emails. Queries containing these entities are processed by a newly created virtual identity.

As stated above, the dissociation permits the user to have some level of privacy. The more virtual identities, the best privacy level achieved. At the same, the search engine will give some personalization to each virtual identity.

Virtual identities are processed by DisPA on the assumption that the search engine only uses cookies to identify the users. Additional tools to avoid e.g. device fingerprinting are not currently implemented. Recall that device fingerprinting [25,26] is an easy path to reidentification.

4.3   Private Information Retrieval

Private Information Retrieval (PIR) studies how a user can retrieve an object from a database (or a search engine) in such a way that the database server is not aware of the object being retrieved. This problem can be formalized as follows.

Definition 4.1 Let the database be represented by a binary string of n bits $x = x_1, \ldots, x_n$. Therefore, $x_i \in \{0, 1\}$. Given a value $i \in \{1, \ldots, n\}$, the private information retrieval problem consists of retrieving x_i without the database server knowing which is the bit being retrieved.

The following result states that when there are no constraints on the computation capabilities of the database server, the only way to achieve privacy is to copy the whole database. We call information theoretic PIR this approach in which there are no constraints on the computational capabilities of the server.

Theorem 4.1 [27–29] *Any information theoretic PIR scheme with a single-database with n bits requires $\Omega(n)$ bits of communications.*

Due to this result, two approaches have been considered that relax the conditions that can be found in the previous theorem. One focuses on having multiple copies of the database (information theoretic PIR with several databases). The other focuses on servers with limited computational power (computational PIR).

We give examples of information-theoretic PIR in Sect. 4.3.1 and of computational PIR in Sect. 4.3.2. The literature presents other approaches on both lines.

4.3.1   Information-Theoretic PIR with k Databases

Let us consider procedures for this type of PIR. That is, we consider the problem of accessing the bit i in a database with n bits $x_1, \ldots, x_n$, when there are k copies of this database. First, we formalize the scenario. It consists of k servers each one with a copy of the database. In addition to the value i, the user will use a random vector of length l_{rnd}.

Then, the user will send k queries, one to each server. Each server will return an answer to the query received. Then, the user will combine the k answers into a value that is the bit x_i.

The next definition formalizes this problem.

Definition 4.2 [28] Let us consider the following functions to query k servers, each one with a copy of a database of length n.

- k query functions $Q_1, \ldots, Q_k : \{1, \ldots, n\} \times \{0, 1\}^{l_{rnd}} \rightarrow \{0, 1\}^{l_q}$,
- k answer functions $A_1, \ldots, A_k : \{0, 1\}^n \times \{0, 1\}^{l_q} \rightarrow \{0, 1\}^{l_a}$, and
- a reconstruction function $R : \{1, \ldots, n\} \times \{0, 1\}^{l_{rnd}} \times (\{0, 1\}^{l_a})^k \rightarrow \{0, 1\}$.

This construction is a theoretic private information retrieval (PIR) scheme for the database of length n if the functions satisfy the following two properties.

- **Correctness.** For every $x \in \{0,1\}^n$, $i \in \{1,\ldots,n\}$, and $r \in \{0,1\}^{l_{rnd}}$

$$R(i, r, A_1(x, Q_1(i, r)), \ldots, A_k(x, Q_k(i, r))) = x_i.$$

- **Privacy.** For every $i, j \in \{1,\ldots,n\}$, $s \in \{1,\ldots,k\}$, and $q \in \{0,1\}^{l_q}$,

$$Pr(Q_s(i, r) = q) = Pr(Q_s(j, r) = q)$$

where the probabilities are taken over uniformly chosen $r \in \{0,1\}^{l_{rnd}}$.

We will describe below two schemes that are theoretic private PIR. The first one is for $k = 2$ and the other for $k \geq 2$. Both are given in [28].

Let $k = 2$ and let SDB_1 and SDB_2 denote the two database servers. Let i be the index of the bit the user is interested in. Then, the scheme is as follows.

- **Step 1.** The user selects a random set $S \subset \{1,\ldots,n\}$ where each index $j \in \{1,\ldots,n\}$ is selected with probability $1/2$.
- **Step 2.** The user sends $Q_1 = S$ to server SDB_1 and $Q_2 = S \boxplus i$ to server SDB_2 where $S \boxplus i$ is defined as

$$S \boxplus i = \begin{cases} S \cup \{i\} & \text{if } i \notin S \\ S \setminus \{i\} & \text{if } i \in S \end{cases}$$

That is, $Q_1 = b_1^1 \ldots b_n^1$ where $b_i^1 = 1$ if and only if $i \in S$ and $Q_2 = b_1^2 \ldots b_n^2$ where $b_i^2 = 1$ if and only if $i \in S \boxplus i$.
- **Step 3.** The server SDB_1 sends an exclusive-or of the bits in S and the server SDB_2 sends an exclusive or of the bits in $S \boxplus i$. That is,

$$A_1 = \bigoplus_{j \in S} x_j \tag{4.4}$$

$$A_2 = \bigoplus_{j \in S \boxplus i} x_j. \tag{4.5}$$

- **Step 4.** The user computes the exclusive or of the bits received. That is,

$$R = A_1 \oplus A_2.$$

We can prove the correctness of the procedure observing that R results into x_i. Note that

$$R = A_1 \oplus A_2 = \bigoplus_{j \in S} x_j \bigoplus_{j \in S \boxplus i} x_j = \bigoplus_{j \in S \setminus \{i\}} (x_j \oplus x_j) \oplus x_i = x_i.$$

In addition, the servers do not get any information about the index desired by the user. This is so because each server receives a uniformly distributed subset of $\{1, \ldots, n\}$.

As a final remark note that this solution requires the servers to send a single bit (i.e., $l_a = 1$). Note, however, that the number of bits sent by the user has length n.

The second scheme is also for calculating the bit i in $\{1, \ldots, n\}$ from a set of k servers. The scheme is valid for $k \geq 2$ when k can be expressed as 2^d for a given $d \geq 1$. Here we also assume that $n = l^d$ for a given l.

The scheme requires first some definitions. We need to embed the database x in a d-dimensional cube. This is done associating each $j \in \{1, \ldots, n\}$ a tuple $(j_1, \ldots, j_d) \in [1, \ldots, l]^d$ in a natural way. We call this function ϕ. For example, if $\phi(j) = (j_1, \ldots, j_d)$, then $j = \sum_{i=1}^{d}(j_i - 1)l^{i-1}$.

Then, we make a one-to-one correspondence between the $k = 2^d$ servers and the string in $\{0, 1\}^d$.

- **Step 1**. The user calculates $(i_1, \ldots, i_d) = \phi(i)$ for the given i.
- **Step 2**. The user chooses uniformly and independently d random subsets

$$S_1^0, S_2^0, \ldots, S_d^0 \subseteq \{1, \ldots, l\}.$$

- **Step 3**. The user computes $S_i^1 = S_i^0 \boxplus i_i$ for $i = 1, \ldots, d$.
- **Step 4**. There are $k = 2^d$ servers with names $\{0, 1\}^d$. The user sends a query to each server. Let us consider the server named $\alpha = (\alpha_1, \ldots, \alpha_d)$. Note that α_i is either zero or one. The user sends to this server SDB_α the subsets

$$S_1^{\alpha_1}, S_2^{\alpha_2}, \ldots, S_d^{\alpha_d}.$$

- **Step 5**. The product of the sets $S_1^{\alpha_1}, S_2^{\alpha_2}, \ldots, S_d^{\alpha_d}$ define a subcube of the d-dimensional cube associated to x.
  Each server SDB_α returns to the user the exclusive-or of the bits in the subcube queried. That is,

$$A_\alpha = \bigoplus_{j_1 \in S_1^{\alpha_1}, \ldots, j_d \in S_d^{\alpha_d}} x_{j_1, \ldots, j_d},$$

or, equivalently

$$A_\alpha = \bigoplus_{(j_1, \ldots, j_d) \in S_1^{\alpha_1} \times \cdots \times S_d^{\alpha_d}} x_{j_1, \ldots, j_d}.$$

- **Step 6**. The user receives one bit from each server and computes the exclusive-or of these bits. There are $k = 2^d$ bits to combine. That is:

$$R = \oplus_\alpha A_\alpha.$$

Reference [28] proves the correctness and privacy of this approach. The number of bits sent is

$$k \cdot (d \cdot l + 1) = 2^d (d \cdot n^{1/d} + 1).$$

Another more balanced scheme is also presented in [28]. That is, while here the user sends $d \cdot n^{1/d}$ bits, it receives only one from each server. The balanced schemes are to reduce the number of bits sent increasing the number of bits received.

4.3.2 Computational PIR

Computational Private Information Retrieval (cPIR) ensures privacy on the basis that databases have restricted computational capabilities. More specifically, privacy is on the basis that databases are restricted to perform only polynomial-time computations. Initial results in this area were presented in [30,31]. Some methods required several copies of the databases, others can be applied with only one database. That is, replication is not needed.

In this section we present a scheme for computational PIR without replication of the data. The scheme, from [29], is based on the intractability of the quadratic residuosity problem. Let us first review this problem. We need the following definition about coprimes.

Definition 4.3 Let N be a natural number, then Z_N^* is the set of numbers defined by:

$$Z_N^* = \{x | 1 \le x \le N, gcd(N, x) = 1\}.$$

That is, Z_N^* is the set of numbers coprime with N.

Now, we define the quadratic residue modulo N.

Definition 4.4 An integer y is a quadratic residue modulo N if there is an integer x such that

$$x^2 \equiv y \pmod{N}.$$

We define a predicate for this problem.

Definition 4.5 The quadratic residuosity predicate is defined as follows:

$$Q_N(y) = \begin{cases} 0 \text{ if there exists a number } w \text{ in } Z_N^* \text{ such that } w^2 \equiv y \pmod{N} \\ 1 \text{ otherwise} \end{cases}$$

$$(4.6)$$

That is, $Q_N(y)$ is zero when y is the quadratic residue modulo N of a number w coprime with N. If such number w does not exist, then $Q_N(y) = 1$. In addition, we say that y is QR if it is a quadratic residue ($Q_N(y) = 0$) and that y is QNR if it is a quadratic non-residue ($Q_N(y) = 1$).

Definition 4.6 Given an integer a and an odd primer number p, the Legendre symbol is defined by:

$$\left(\frac{a}{p}\right) = \begin{cases} 0 & \text{if } a \equiv 0 \pmod{p} \\ 1 & \text{if } a \not\equiv 0 \pmod{p} \text{ and } a \text{ is a quadratic residue modulo } p \\ -1 & \text{if } a \text{ is a quadratic non-residue modulo } p \end{cases}$$

(4.7)

The following properties hold for the Legendre symbol.

Proposition 4.1 *Let p be an odd prime number, let a and b be integers. Then,*

- *the following equality holds*

$$\left(\frac{ab}{p}\right) = \left(\frac{a}{p}\right)\left(\frac{b}{p}\right),$$

- *when $a \equiv b (mod p)$, the following holds*

$$\left(\frac{a}{p}\right) = \left(\frac{b}{p}\right).$$

Definition 4.7 Let a be an integer and n be a positive odd integer decomposable in terms of prime numbers $p_1, \ldots, p_k$ as follows $n = p_1^{a_1} \cdots\cdots p_k^{a_k}$. Then, the Jacobi symbol is defined in terms of the Legendre symbol as follows

$$\left(\frac{a}{n}\right) = \left(\frac{a}{p_1}\right)^{a_1} \cdot \left(\frac{a}{p_2}\right)^{a_2} \cdots\cdots \left(\frac{a}{p_k}\right)^{a_k}.$$

The Jacobi symbol generalizes the Legendre symbol as n can now be any positive odd integer, and when n is an odd prime both Jacobi and Legendre symbols are equal.

There are a few computational aspects of interest related to the computation of $Q_N(y)$ and the Legendre and Jacobi symbols.

1. Let H_k be the set of integers that are the product of two primes of length $k/2$ bits. That is,

$$H_k = \{N | N = p_1 \cdot p_2 \text{ where } p_1, p_2 \text{ are } k/2 - \text{ bit primes}\}.$$

Then, if $N \in H_k$ and its factorization is known, the computation of $Q_N(y)$ can be done in $O(|N|^3)$.

In contrast, when the factorization of N is not known, the computation of $Q_N(y)$ is intractable (see e.g. [29]).

2. For all N (even for $N \in H_k$, and without knowing its factorization), the Jacobi symbol

$$\left(\frac{a}{N}\right)$$

can be computed in polynomial time of $|N|$.
3. For all $N \in H_k$, given an integer y, the computation of the Jacobi symbol

$$\left(\frac{y}{N}\right)$$

can either be $+1$ or -1.

- If it is -1, we know that y is a quadratic non-residue.
- In contrast, if it is $+1$ the integer y can be either quadratic residue or quadratic non-residue. Not only it can be either QR and QNR, but the sets of QR and QNR for a given N are of the same size.
  We will denote the set of integers for which this value is one by Z_N^+. That is,

$$Z_N^+ = \left\{ y \in X_N^* \mid \left(\frac{y}{N}\right) = 1 \right\}$$

4. For all x, y in Z_N^+ it holds that

$$Q_N(xy) = Q_N(x) \oplus Q_N(y),$$

where $\oplus$ denotes the exclusive or. In other words, the product of two integers is QNR if and only if one of them is QNR.

As a summary we define the quadratic residue problem and review the complexity of solving this problem.

Definition 4.8 Given integers a and N with N the product of two different primes p_1 and p_2 and such that

$$\left(\frac{a}{N}\right) = 1$$

the quadratic residuosity problem is to determine if a is quadratic residue modulo N or not. That is, compute $Q_N(a)$.

For N in H_k and y in Z_N^+, we have that if we know the factorization of N the computation of $Q_N(y)$ is polynomial but when the factorization is not known the computation of $Q_N(y)$ is intractable.

The algorithm below (see Sect. 3 in [29]) for computational PIR is based on this fact. The algorithm considers a server SDB with a database. We represent the database as a matrix M (of bits) with s rows and t columns.

The goal of the algorithm is that the user retrieves the bit at the position (a, b) of a matrix M. That is, the content of row a and column b. We denote this value by $M_{a,b}$.

- **Step 1.** The user selects two random primes of $k/2$-bits each and multiplies them. Let N be the result of this multiplication. Therefore, $N \in H_k$. The user sends N to the server SDB. The factorization is kept secret.
- **Step 2.** The user chooses uniformly at random t numbers $y_1, \ldots, y_t \in Z_N^{+1}$, one for each column. The one of column b, that is y_b, is a QNR and all the others, y_j for $j \neq b$, are QR. The user sends these t numbers to the server. Note that the total number of bits sent corresponds to $t \cdot k$.
- **Step 3.** The database server SDB computes a number z_r for each row $r = 1, \ldots, s$ of the matrix as follows.

  – Compute $w_{r,j}$ for each $j = 1, \ldots, t$ as follows

  $$w_{r,j} = \begin{cases} y_j^2 & \text{if } M_{r,j} = 0 \\ y_j & \text{if } M_{r,j} = 1. \end{cases}$$

  – Compute z_r using $w_{r,j}$ as follows

  $$z_r = \prod_{j=1}^{t} w_{r,j}.$$

Note that this procedure computes a value $w_{r,j}$ for each position (r, j) in the matrix M and then aggregates all the values of the rth row into z_r.

The definition of $w_{r,j}$ is such that $w_{r,j}$ is always a QR when $j \neq b$. This is so because y_j is a QR for $j \neq b$. In contrast, when $j = b$, we have that $w_{r,j}$ is QR if and only if $M_{r,j} = 0$. This is so because when $M_{r,b} = 1$ then $w_{r,b} = y_b$ which is QNR and when $M_{r,b} = 0$ then $w_{r,b} = y_b^2$ which is QR. Because of that, z_r is a QR if and only if $M_{r,b} = 0$ (this follows from the condition above that for x, y in Z_N^+, $Q_N(xy)$ is the exclusive or of $Q_N(x)$ and $Q_N(y)$). This computation is represented in Fig. 4.5.

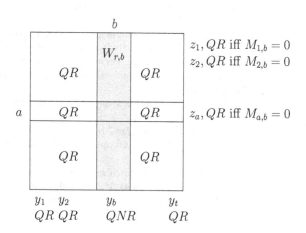

Fig. 4.5 Computation of values z_r for the scheme in [29] for computational PIR

- **Step 4**. The server SDB sends $z_1, \ldots, z_s$ to the user. The total number of bits sent is $s \cdot k$.
- **Step 5**. The user considers only the number z_a. Recall that a is the row of M which contains the bit the user is interested in. This number is a QR if and only if $M_{a,b} = 0$ (and QNR otherwise).

  Since the user knows the factorization of the number N, it can check whether z_a is a QR. If it is QR then the user infers that $M_{a,b} = 0$, otherwise $M_{a,b} = 1$.

  As the server does not have the factorization of N, determining whether z_a is QR or QNR is intractable.

4.3.3   Other Contexts

To finish this chapter we want to mention the application of user's privacy into two other contexts: location based services and online social networks. Reference [32] developed an approach for location based services where users ensure their privacy by swapping trajectories. It is somehow related to data swapping and because of that we mention this work again in Sect. 6.1.1. Data is expected to be processed in a centralized way, but individuals preprocess the data to ensure their privacy.

Online social networks are typically centralized. Nevertheless, there are alternative ways of building them from a decentralized perspective. These decentralized online social networks are based on P2P protocols and are constructed to provide user's privacy. See e.g. [33] and PeerSoN [34]. These networks need to include mechanisms so that functionalities satisfy appropriate levels of privacy (see e.g. [35] on event invitations in distributed OSNs).

References

1. Edman, M., Yener, B.: On anonymity in an electronic society: a survey of anonymous communication systems. ACM Comput. Surv. **42**, 1–5 (2009)
2. Pfitzmann, A., Hansen, M.: A terminology for talking about privacy by data minimization: anonymity, unlinkability, undetectability, unobservability, pseudonymity, and identity management. http://dud.inf.tu-dresden.de/literatur/Anon_Terminology_v0.34.pdf (2010)
3. Chaum, D.L.: Untraceable electronic mail, return addresses, and digital pseudonyms. Commun. ACM **24**(2), 84–88 (1981)
4. Serjantov, A.: On the anonymity of anonymity systems. Technical report, Computer laboratory, University of Cambridge (2004)
5. http://mixminion.net/. Accessed Jan 2017
6. Reiter, M., Rubin, A.: Crowds: anonymity for web transactions. ACM Trans. Inf. Syst. Secur. **1**(1), 66–92 (1998)
7. Reed, M.G., Syverson, P.F., Goldschlag, D.M.: Anonymous connections and onion routing. IEEE J. Sel. Areas Commun. **16**(4), 482–494 (1998)
8. Dingledine, R., Mathewson, N., Syverson, P.: TOR: the second-generation onion router. In: 13th USENIX Security Symposium (2004)

9. https://www.torproject.org/. Accessed Jan 2017
10. Chaum, D.: The dining cryptographers problem: unconditional sender and recipient untraceability. J. Crypt. **1**, 65–75 (1985)
11. Domingo-Ferrer, J., Bras-Amorós, M., Wu, Q., Manjón, J.: User-private information retrieval based on a peer-to-peer community. Data Knowl. Eng. **68**(11), 1237–1252 (2009)
12. Stokes, K., Bras-Amorós, M.: On query self-submission in peer-to-peer user-private information retrieval. In: Proceedings of 4th PAIS 2011 (2011)
13. Stokes, K., Bras-Amorós, M.: Optimal configurations for peer-to-peer user-private information retrieval. Comput. Math. Appl. **59**(4), 1568–1577 (2010)
14. Stokes, K., Bras-Amorós, M.: A survey on the use of combinatorial configurations for anonymous database search. In: Navarro-Arribas, G., Torra, V. (eds.) Advanced Research in Data Privacy. Springer, Cham (2015)
15. Stokes, K., Farràs, O.: Linear spaces and transversal designs: k-anonymous combinatorial configurations for anonymous database search. Des. Codes Cryptogr. **71**, 503–524 (2014)
16. Howe, D.C., Nissenbaum, H.: TrackMeNot: resisting surveillance in web search. In: Kerr, I., Steeves, V., Lucock, C. (eds.) Lessons from the Identity Trail: Anonymity, Privacy, and Identity in a Networked Society, pp. 417–436. Oxford University Press, Oxford (2009)
17. Domingo-Ferrer, J., Solanas, A., Castella-Roca, J.: $h(k)$-private information retrieval from privacy-uncooperative queryable databases. Online Inf. Rev. **33**(4), 720–744 (2009)
18. Peddinti, S.T., Saxena, N.: On the privacy of web search based on query obfuscation: a case study of TrackMeNot. In: Proceedings of the Privacy Enhancing Technologies. LNCS, vol. 6205, pp. 19–37 (2010)
19. Juárez, M., Torra, V.: A self-adaptive classification for the dissociating privacy agent. In: Proceedings of PST 2013, pp. 44–50 (2013)
20. Juárez, M., Torra, V.: Toward a privacy agent for information retrieval. Int. J. Intell. Syst. **28**(6), 606–622 (2013)
21. Binder, J., Howes, A., Sutcliffe, A.: The problem of conflicting social spheres: effects of network structure on experienced tension in social network sites. In: Proceedings of CHI 2009 (2009)
22. Nissenbaum, H.: Privacy as contextual integrity. Washington Law Rev. **79**, 119–158 (2004)
23. Juàrez, M., Torra, V.: DisPA: an intelligent agent for private web search. In: Navarro-Arribas, G., Torra, V. (eds.) Advanced Research on Data Privacy, pp. 389–405. Springer, Cham (2015)
24. ODP: Open directory project. http://www.dmoz.org/. Accessed Jan 2017
25. Acar, G., Juarez, M., Nikiforakis, N., Diaz, C., Gürses, S., Piessens, F., Preneel, B. FPDetective: dusting the web for fingerprinters. In: Proceedings of the ACM Conference on Computer and Communications Security (CCS), pp. 1129–1140 (2013)
26. Eckersley, P.: How unique is your browser? In: Proceedings 10th Privacy Enhancing Technologies Symposium (PETS), pp. 1–17 (2010)
27. Chor, B., Goldreich, O., Kushilevitz, E., Sudan, M.: Private information retrieval. In: Proceedings of the IEEE Conference on Foundations of Computer Science, pp. 41–50 (1995)
28. Chor, B., Goldreich, O., Kushilevitz, E., Sudan, M.: Private information retrieval. J. ACM **45**(6), 965–982 (1999)
29. Kushilevitz, E., Ostrovsky, R.: Replication is not needed: single database, computationally-private information retrieval. In: Proceedings of the 38th Annual Symposium on Foundations of Computer Science, pp. 364–373 (1997)
30. Chor, B., Gilboa, N.: Computationally private information retrieval. In: Proceedings of the 29th STOC, pp. 304–313 (1997)
31. Ostrovsky, R., Shoup, V.: Private information storage. In: Proceedings of the 29th STOC, pp. 294–303 (1997)
32. Gidófalvi, G.: Spatio-temporal data mining for location-based services. Ph.D. dissertation (2007)

33. Datta, A., Buchegger, S., Vu, L.-H., Strufe, T., Rzadca, K.: Decentralized online social networks. In: Handbook of Social Network Technologies, pp. 349–378 (2010)
34. Buchegger, S., Schiöberg, D., Vu, L.-H., Datta, A.: PeerSoN: P2P social networking: early experiences and insights. In: Proceedings of SNS 2009, pp. 46–52 (2009)
35. Rodríguez-Cano, G., Greschbach, B., Buchegger, S.: Event invitations in privacy-preserving DOSNs - formalization and protocol design. IFIP Advances in Information and Communication Technology, vol. 457, pp. 185–200 (2015)

Privacy Models and Disclosure Risk Measures

5

Elàstics blaus subjectats amb candaus
porta el meu enamorat i el barret de costat,
de color verd, que és el que em perd.

I porta un gec, catacric, catacrec,
un gec d'astracan pelut, ribetat de vellut,
i a l'armilla hi duu cigrons per botons.

Joan Casas i Vila, El vestir d'en Pasqual, 1920s.

This chapter is devoted to the study of different approaches to define and evaluate the risk of disclosure. As stated in Chap. 1 (Sect. 1.3.3) there are two types of disclosure: attribute disclosure and identity disclosure. In order to evaluate masking methods, models and measures for both attribute and identity disclosure have been developed.

5.1 Definition and Controversies

As already pointed out in Sect. 1.1 (see e.g. [1,2]), there is some discussion on the importance of reidentification and identity disclosure in real data and whether the risk due to reidentification techniques should prevent data dissemination. Discussion is based on the following:

- only a few cases of reidentifications are known, and most privacy breaches are due to causes (i.e., security breaches) other than to a bad use of data privacy techniques;

© Springer International Publishing AG 2017
V. Torra, *Data Privacy: Foundations, New Developments and the Big Data Challenge*, Studies in Big Data 28,
DOI 10.1007/978-3-319-57358-8_5

- reidentification algorithms are not straightforward to use and the gain an intruder can obtain from analysing a data set may be very low.

In addition to this, as [3] underlines, reidentification can only be done when the adversary has some additional information on the person being reidentified. So, the release of a data set with a certain disclosure risk does not imply that a bunch of intruders will have the time and interest to try to deindentify the data set.

Similarly, we have discussed in Sect. 1.3.3 that attribute disclosure is in some extent natural. Otherwise, the interest of the published data may be extremely limited.

Arguments stating that risk should not be overestimated can be seen as at one extreme of the spectrum of controversies. On the other, we have those that state that privacy is impossible because disclosure risk cannot be avoided unless data become useless. This position has been supported by e.g. [4,5]. The later underline the difficulties of protecting data correctly, the data releases that lead to disclosure due to insufficient protection (e.g. the AOL and the Netflix cases) and the fact that high dimensional data may need a large perturbation to make them safe.

Reference [4] points to the difficulty of avoiding disclosure discussing the risk of deidentified credit card data. This paper was replied in [6,7] (which lead to a subsequent answer from the authors in [8,9], respectively). Replies point out that uniqueness in a sample does not imply uniqueness in the population (we discuss uniqueness in Sect. 5.3.3) and that data masking methods used in the experiments were not effective (appropriate).

All these arguments on the difficulties of data privacy are partially true but this does not mean that privacy is impossible. The difficulty and success of data protection will depend on the type of data we have and the level of protection we want to achieve. The selection of an appropriate method is essential to have a protected data set that does not lead to disclosure.

All these discussions only point out the importance of defining correctly disclosure risk, and the need of accurate ways of measuring disclosure risk in order to be able to make informed decisions. Then, the values of these measures have to be taken and used by the decisor within the appropriate context.

In addition to the use of these measures by a decisor in order to know if data can be released or not, these measures are also used to evaluate and compare different masking methods as well as different instantiations of the same masking method using different parameters. So, the utility of the measures is two fold: make decisions on data releases and compare masking methods.

For further discussion on the use of reidentification for disclosure risk assessment, consider e.g. [10] which advocates for the evaluation of the risk of reidentification.

One of the first detailed discussions on disclosure (statistical disclosure) is due to Dalenius [11]. He points out the different sets of data that are relevant when evaluating disclosure risk, and presents a classification of the types of disclosure. Figure 5.1 gives an overview of Dalenius dimensions for disclosure risk assessment.

With respect to the sets, he distinguishes between the total population, the subset of this population that is accessible in a certain frame, and the sample survey. These sets are illustrated in the following example.

Dalenius (1977) classifies disclosure according to the following dimensions. The definition uses D to denote a characteristic, and D_K to denote the characteristic for the kth record. We will use A as used elsewhere in this text to denote an original attribute, and $A(x_k)$ to denote the value of A for the kth record x_k. We will use A' to denote the attribute of a protected file, and A^* to denote the estimation of A using the published dataset. In most cases $A^* = A'$ if A' is in the protected dataset, but in some cases, this is not the case. Then, inference procedures can be used and then the equality does not apply.

1. Kinds of statistics released. Two cases are considered: statistics for sets of objects (macro-statistics) and statistics for individual objects (micro-statistics). The former corresponds to tabular data and the latter to microdata.
2. The measurement scale used to express the data. A is used to denote the characteristic. Then, two main cases are considered: scales yielding attribute/counts (binary data i.e., A is either 0 or 1 for record x_k) and yielding magnitudes (Dalenius considers A measured in a ratio scale, see the following references for details on scales and measurement theory [12, 13, 14, 15]).
3. Accessibility of disclosure. Direct and indirect disclosure are distinguished. The former corresponds to the case that A is included in the published dataset, and the latter to the case that A is computed from the published dataset.
4. Accuracy of disclosure. Exact and approximate disclosure are distinguished. We have *exact disclosure* when the published data corresponds to the original data. In contrast, we have *approximate disclosure* when this is not the case and there is some imprecision or uncertainty in the disclosure. Two types of approximate disclosure are considered: approximation of the true value by means of an interval $[A_L, A_U]$ (i.e., $A_L \leq A^*(x_k) \leq A_U$) and approximation in terms of a category (i.e., a boolean value $A^*(x_k) \in \{0,1\}$). Dalenius further considers that an approximation can be *certain* or *uncertain*. The former corresponds to the case that the object belongs to the interval or category involved, the latter when this is not always the case and, therefore, there is some uncertainty. Probabilities are considered to model this latter case.
5. External versus internal disclosure. We have external disclosure when disclosure for record x_k takes place without information about other objects x_j in the released dataset. Internal disclosure takes place when disclosure for x_k takes advantage of information about other records x_j with $j \neq k$. We might have both external and internal disclosure. Let $A^*(x_k)$ be the best approximation of $A(x_k)$ obtained using only data from x_k, and $A^{**}(x_k)$ the best approximation obtained using data from other records x_j with $j \neq k$. Then, if $A^{**}(x_k)$ is a better approximation to $A(x_k)$ than $A^*(x_k)$, we have internal disclosure. Note that we have interval disclosure when a collusion of individuals use their information to discover information about another one.
6. The disclosing entities. S-based disclosure and $S \times E$-based disclosure. Here E, which denotes the term extra-objective data, corresponds to any kind of additional data. These data are not part of the objective of the survey. Naturally, we have S-based disclosure when only the information of the released data set is used, and $S \times E$-based disclosure when additional information is used to have better approximations A^* of A.

Fig. 5.1 Statistical disclosure according to Dalenius [11]

Example 5.1 [11] Let us consider the release of data from a sample of Swedish citizens. In this case, the following sets are of interest.

1. $\{O\}_T$ denotes the total population: *all Swedish citizens.*
2. $\{O\}_F$ denotes the accessible population: *all Swedish citizens living in Sweden.*
3. $\{O\}_{F,s}$ denotes the individuals selected in the sample: *all Swedish citizens living in Sweden selected in the sample.*

Then, let $x \in \{O\}_T$. We may have $x \in \{O\}_F$ or $x \notin \{O\}_F$. Let A be a characteristic (an attribute), either one of the ones in the survey (or data set) or another one. Thus $A(x)$ is the value of this characteristic for x. Dalenius defines disclosure as follows.

Definition 5.1 [11] (p. 433) If the release of the statistics S makes it possible to determine the value $A(x)$ more accurately than is possible without access to S, a disclosure has taken place.

This definition corresponds to what we have defined as attribute disclosure.

Dalenius, in a latter paper [16], also considered what we call identity disclosure. He considered the case that a set of variables makes a record unique (he calls a certain combination of values/variables a P-set of data) or almost unique (less than k individuals with $k = 2$ or 3 for a given combination of values/variables). In the latter case, Dalenius discusses about the possibility that a collusion of individuals permit them the reidentification of another one.

5.1.1 A Boolean or Measurable Condition

Disclosure can be seen as a Boolean condition or as a measurable one. Differential privacy and k-anonymity are two example of Boolean conditions. They define a condition that when satisfied implies that our requirements of privacy are satisfied and, thus, disclosure risk does not take place. We have seen differential privacy in Sect. 3.4.1. In differential privacy, ε corresponds to the privacy level. Given ε, we can check whether privacy is guaranteed. We review k-anonymity in Sect. 5.8. As we will see, it is related to reidentification.

- Types of disclosure according to what is disclosed.

  1. Attribute disclosure (Section 5.2)
  2. Identity disclosure (Section 5.3)

- Privacy models and measures of disclosure.

  1. As a measurable condition
     a. Uniqueness (Section 5.3.3)
     b. Reidentification (Section 5.3.4)
        i. Data integration: schema and data matching
        ii. Record linkage algorithms: distance-based and probabilistic RL
        iii. Generic vs. specific record linkage algorithms
  2. As a Boolean condition
     a. Differential privacy (Section 3.4.1)
     b. k-Anonymity (Section 5.8)
     c. Interval disclosure (Section 5.2.1)

Fig. 5.2 Outline of privacy models and disclosure risk measures

The definition of disclosure as a Boolean condition permits us to look for masking methods that minimize information loss for a certain privacy level. Note that this is not the case when risk is a measurable condition. In this latter case, the goal is to find a good balance between information loss and disclosure risk. Therefore, the search for a good masking method corresponds to a multicriteria optimization problem.

Figure 5.2 gives an outline of the different approaches and topics we discuss in this book related to disclosure risk assessment.

5.2 Attribute Disclosure

Attribute disclosure happens when intruders can increase the accuracy of their information with respect to the value of an attribute for a particular individual.

This is a very general definition and, because of that, all functions that measure the divergence between an original value and a protected value, or between the true value and the value that can be inferred from the protected data can be used for measuring attribute disclosure.

Among the existing measures for attribute disclosure we describe here one approach for numerical data and another for nominal (categorical) data.

5.2.1 Attribute Disclosure for a Numerical Variable

The attribute is ranked and a rank interval is defined around the value the attribute takes on each record. The ranks of values within the interval for an attribute around record r should differ less than p percent of the total number of records and the rank in the center of the interval should correspond to the value of the attribute in record r. Then, the proportion of original values that fall into the interval centered around their corresponding protected value is a measure of disclosure risk. A 100% proportion means that an attacker is completely sure that the original value lies in the interval around the protected value (interval disclosure). Algorithm 12 describes interval disclosure for a given attribute and record. We call this method rank-based interval disclosure.

Interval disclosure is a measure proposed in [17, 18] for numerical data. A variation was introduced in [19] in which the interval is based on a percentage of the standard deviation of the variable. This definition is called *SDID*, for standard deviation-based interval disclosure. See Algorithm 13.

The R package sdcMicro [20] includes a multivariate version of this definition. The function *dRiskRMD* defines the center of the interval in terms of the original value and checks whether the masked value is within a certain distance computed using the robust Mahalanobis distance. Note that this is to replace $|V(x) - V'(x)|$ by an appropriate distance in Algorithm 13.

Algorithm 12: Rank-based interval disclosure: $rid(X, V, V', x, p)$.

Data: X: Original file; V: Original attribute; V': Masked attribute; x: record; p: percentage

Result: Attribute disclosure for attribute V' of record x

1 $R(V) :=$ Rank data for attribute V' ;
2 $i :=$ position of $V'(x)$ in $R(V)$;
3 $w := p \cdot |X|/2/100$ (width of the interval) ;
4 $I(x) = [R[\min(i - w, 0)], R[\max(i + w, |X| - 1)]]$ (definition of the interval for record x);
5 $rid := V(x) \in I(x)$;
6 **return** rid;

Algorithm   13:   Standard   deviation-based   interval   disclosure: $sdid(X, V, V', x, p)$.

Data: X: Original file; V: Original attribute; V': Masked attribute; x: record; p: percentage

Result: Attribute disclosure for attribute V' of record x

1 $sd(V) :=$ standard deviation of V ;
2 $sdid := |V(x) - V'(x)| \leq p \cdot sd(V)/100$;
3 **return** $sdid$;

5.2.2 Attribute Disclosure for a Categorical Variable

When the goal is to evaluate attribute disclosure for an ordinal attribute we can apply the method defined in [21]. This method computes attribute disclosure risk for a given attribute in terms of a particular model or classifier for this attribute constructed from the released data. The percentage of correct predictions given by the classifier is a measure of the risk.

Let us describe this approach in more detail. We use X to denote the original file, $X' := \rho(X)$ to denote the masked file using protection method ρ, and V to denote the nominal attribute of X under consideration. That is, the attribute for which we want to compute attribute disclosure.

First, we need to define the particular model or classifier we will use for the given attribute. As a simple example, we consider the case that the adversary uses the most similar record of the masked data set to guess the value of the attribute *illness*. That is, we apply the nearest neighbor model (See Sect. 2.2.1). Then, for a given record x the adversary selects the most similar record in the released data set X'. Let x' be this record of X'. Therefore, $x' = \arg\min_{x_i \in X'} s(x, x_i)$ for a given similarity function s. Then the adversary uses the attribute illness of x' to infer the illness of x. The percentage of times that this approach succeeds can be used to measure attribute disclosure risk.

In general, the attribute disclosure for attribute V is estimated for a pair of files X and X' as the proportion of correct answers of the records in X using the information in X' using the algorithm A.

In order to estimate this proportion, the approach described in [21] is inspired in k-fold cross validation. Using the notation in Sect. 2.2.3, we have k training sets C_i^{Tr}

for $i = 1, \ldots, k$. Then, we build k different models $M_{C_i^{Tr}, A}$ for $i = 1, \ldots, k$ each one in terms of the set C_i^{Tr} using algorithm A. Finally, attribute disclosure corresponds to the proportion of correct answers. This is computed taking into account for each x in C_i the outcome of the model $M_{C_i^{Tr}, A}$ for such x and comparing with the true value of $V(x)$ that can be found in X.

Algorithm 15 details this process. For the computation of the value $M_{C_i^{Tr}, A}(x)$ Algorithm 14 is used.

In [21], see also [22], attribute disclosure risk was evaluated for three different datasets using the following classifiers available in WEKA [23]: Decision trees, naive Bayes, k-nearest neighbors, and support vector machines. Datasets were protected using the following masking methods: additive and multiplicative noise, rank swapping, and microaggregation. The experiments showed that these definitions lead to high values of attribute disclosure for the protection methods selected.

5.2.2.1 Other Measures and Discussion

The literature presents other measures for evaluating disclosure. See e.g. the use of mutual information in [24] and statistical dependence ($\theta = 1 - \rho^2$) in [25]. These measures can be seen as for attribute disclosure. Note that the measures for attribute disclosure can also be related to measures for information loss.

It is important to recall that attribute disclosure may be acceptable when no identity disclosure takes place. In particular, from a machine learning and statistics perspective, we can state that high values of attribute disclosure imply that classifiers are quite robust to the noise introduced by the masking method. So, when no identity disclosure takes place, this can be seen as a positive result. A more detailed discussion from this perspective is given in [22]. Reference [22] compares these results with additional experiments on identity disclosure and information loss showing that some methods have low identity disclosure risk with high accuracy (high attribute disclosure!).

Algorithm 14: Determination of the class of example x according to the algorithm A and the training set C_{tr}: $M_{C_{tr}, A}(x)$.

Data: C_{tr}: training set; A: algorithm to build a classification model; x: an example
Result: class for x
1 Take the dataset C_{tr} as input and construct a model of the variable V using the algorithm A. Let this model be denoted by $M_{C_{tr}, A}$;
2 $V'(x) := M_{C_{tr}, A}(x)$. That is, the application of the classifier $M_{C_{tr}, A}$ to the data x ;
3 **return** $V'(x)$;

Algorithm 15: Attribute disclosure for attribute V and data sets X, X' according to the algorithm A: $adr_{M,X',X}(A)$.

Data: X: original set; X': masked data set; A: algorithm to build a classification model
Result: attribute disclosure risk
1 Define C_i^{Tr} as in Sect. 2.2.3 (cross-validation);
2 Compute attribute disclosure risk as follows

$$adr_{M,C,X} := \frac{\sum_{i=1}^{k} |\{x | x \in C_i \text{ and } M_{C_i^{Tr},A}(x) = V(x)\}|}{|X|}.$$

Note that here $V(x)$ refers to the true value of attribute V of x (i.e., according to the original data set) ;
3 **return** $adr_{M,C,X}$;

5.3 Identity Disclosure

Identity disclosure takes place when intruders are able to correctly link some of their information on an individual (or entity) with the data in the protected set corresponding to that individual. We can define disclosure risk measures in terms of the number of possible links. In this section we describe the scenario and give an overview of the measures.

5.3.1 An Scenario for Identity Disclosure

We consider a protected data set and an intruder having some partial information about the individuals in the published data set. The protected data set is assumed to be a data file or, more generally, a database. It is usual to consider that intruder's information can be represented in the same way. That is, in terms of a file or database. See e.g. [26,27].

In the typical scenario, the information of the intruder is about some of the attributes published in the protected data set. As always, data are represented by standard files X with the usual structure (r rows (*records*) and k columns (*attributes*)). So, each row contains the values of the attributes for an individual.

Then, the attributes in X can be classified [16,27,28] in the following non-disjoint categories.

- **Identifiers**. These are attributes that *unambiguously* identify the respondent. Examples are passport number, social security number, full name, etc.
- **Quasi-identifiers**. These are attributes that, in combination, can be linked with external information to reidentify some of the respondents. Although a single attribute cannot identify an individual, a subset of them can. For example, the set (*age, zipcode, city*) is a quasi-identifier in some files. Note that although a quasi-identifier is the set of attributes, we will use also this term to denote any of

the attributes in this set. That is, age, zip code, city, birth date, gender, job, etc. will be said to be quasi-identifiers.

- **Confidential**. These are attributes which contain sensitive information on the respondent. For example, salary, religion, political affiliation, health condition, etc. This is usually the information to be studied.
- **Non-confidential**. Attributes not including sensitive information.

Using these categories, an original data set X can be decomposed into $X = id||X_{nc}||X_c$, where id are the identifiers, X_{nc} are the non-confidential quasi-identifier attributes, and X_c are the confidential non-quasi-identifier attributes. We do not consider here non-confidential attributes which are not quasi-identifiers as they are not relevant from a disclosure risk perspective. Confidential quasi-identifiers would be treated as non-confidential quasi-identifiers. We do not mention them explicitly below because the general public will usually not have access to confidential attributes and thus not be able to use them for reidentification.

Let us consider the protected data set X'. X' is obtained from the application of a protection procedure to X. This process takes into account the type of the attributes. It is usual to proceed as follows.[1]

- **Identifiers** id. To avoid disclosure, identifiers are usually removed or encrypted in a preprocessing step. In this way, information cannot be linked to specific respondents.
- **Quasi-identifiers** X_{nc}. They cannot be removed as almost all attributes can be quasi-identifiers. The usual approach to preserve the privacy of the individuals is to apply protection procedures to these attributes. We will use ρ to denote the protection procedure. Therefore, we have $X'_{nc} = \rho(X_{nc})$.
- **Confidential** X_c. These attributes X_c are usually not modified. So, we have $X'_c = X_c$.

Therefore, we have $X' = \rho(X_{nc})||X_c$. Proceeding in this way, we allow third parties to have precise information on confidential data without revealing to whom the confidential data belongs to.

In this scenario we have identity disclosure when intruders, having some information described in terms of a set of records and some quasi-identifiers, can link their information with the published data set. That is, they are able to link their records with the ones in the protected data set. Then, if the links between records are correct,

---

[1]This way of proceeding is already discussed in e.g. [29] (p. 408): "Methods that mask the key variables impede identification of the respondent in the file, and methods that mask the target variables limit what is learned if a match is made. Both approaches may be useful, and in practice a precise classification of variables as keys or targets may be difficult. However, masking of targets is more vulnerable to the trade-off between protection gain and information loss than masking of keys; hence masking of keys seems potentially more fruitful".

intruders have achieved reidentification. In addition to reidentification, they may be able to obtain the right values for the confidential attributes. That is, we will have attribute disclosure.

Figure 5.3 represents this situation. A represents the file with data from the protected data set (i.e., containing records from X') and B represents the file with the records of the intruder. B is usually defined in terms of the original data set X, because it is assumed that the intruder has a subset of X. In general, the number of records owned by the intruder and the number of records in the protected data file will differ.

In the figure, reidentification uses the common quasi-identifiers on both X and X'. They permit to link pairs of records (using record linkage algorithms) from both files. At this point, reidentification is achieved because we link the protected record to an individual (represented by the identifiers). Then, the confidential attribute is linked to the identifiers. At this point attribute disclosure takes place.

Note that this scenario is the one we have discussed in Sect. 1.3.3 about disclosure. See Table 1.1 reproduced here as Table 5.1. We have that when (*city, age, illness*) is published the intruder can find that her friend HYU had a heart attack if she knows that HYU lives in Tarragona and is 58 years old. In this example, (*city, age*) is the quasi-identifier, the illness is the confidential attribute, and the name HYU is the identifier.

Formally, following [26,27,30] and the notation in Fig. 5.3, the intruder is assumed to know the non-confidential quasi-identifiers $X_{nc} = \{a_1, \ldots, a_n\}$ together with

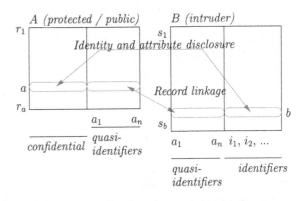

Fig. 5.3 Disclosure risk scenario: attribute and identity disclosure

Respondent	City	Age	Illness
ABD	Barcelona	30	Cancer
COL	Barcelona	30	Cancer
GHE	Tarragona	60	AIDS
CIO	Tarragona	60	AIDS
HYU	Tarragona	58	Heart attack

Table 5.1 Table that permits an adversary to achieve identity disclosure and attribute disclosure. Only attributes City, Age, and Illness are published

the identifiers $Id = \{i_1, i_2, \dots\}$. Then, the linkage is between quasi-identifiers $(a_1, \dots, a_n)$ from the protected data (X'_{nc}) and the same attributes from the intruder (X_{nc}).

In a more general scenario, the database B might have a schema[2] different from the one of database A. That is, the number and type of attributes are different. E.g., while $X_{nc} = \{a_1, \dots, a_n\}$, the database B might have $X'_{nc} = \{a'_1, \dots, a'_m\}$.

Even in the case that there are no common quasi-identifiers in the two files, reidentification may be possible if there is some structural information common in both files. Reference [31] follows this approach. Its use for disclosure risk assessment is described in [32]. Algorithms and methods developed in the field of data integration and schema matching are useful for this type of scenarios. These areas are briefly described in Sects. 5.4.2 and 5.4.3.

5.3.2 Measures for Identity Disclosure

As stated in Sect. 5.1, we can consider disclosure as a Boolean condition or as a measurable one. k-Anonymity, discussed in Sect. 5.8, is an example of Boolean condition for identity disclosure. Reidentification and uniqueness are used for measurable identity disclosure risk. We describe them below.

- **Reidentification**. Risk is defined as an estimation of the number of reidentifications that an intruder may obtain. Given a file with the information of an intruder and the protected file, this estimation is obtained empirically through record linkage algorithms. This approach goes back, at least, to [33,34] (using e.g. the algorithm described in [35]). More recent papers include [26,27,30]. The real anonymity measure we will see in Eq. 6.3 can be seen as inversely proportional to a disclosure risk measure related to reidentification.
- **Uniqueness**. Risk is defined as the probability that rare combinations of attribute values in the protected data set are indeed rare in the original population.
  This approach is typically used when data is protected using sampling [36] (i.e., X' is just a subset of X). Note that with perturbative methods it makes no sense to investigate the probability that a rare combination of protected values is rare in the original data set, because *that* combination is most probably *not found* in the original data set.

In the next section we describe uniqueness in more detail. Reidentification and record linkage are discussed later.

A problem related to identity disclosure is to find minimal uniqueness within a database. SUDA [37] and SUDA2 [38] search for these uniques (called MSU for minimal sample uniques) up to a user-specified size in discrete datasets. This

---

[2]In databases, the schema define the type of attributes, their types and relationships. They roughly correspond to metadata in statistical disclosure control.

limitation to a up to a user-specified size can be linked to the concept of computational anonymity we discuss in Sect. 5.8.3.

5.3.3 Uniqueness

Two types of disclosure risk measures based on uniqueness can be distinguished: file-level and record-level. In the former we have a single measure for the whole file (an average measure) and in the latter we have a measure for each record in the file. We describe both of them below.

- **File-level uniqueness.** Disclosure risk is defined as the probability that a sample unique (SU) is a population unique (PU) [39]. According to [40], this probability can be computed as

$$P(PU|SU) = \frac{P(PU, SU)}{P(SU)} = \frac{\sum_j I(F_j = 1, f_j = 1)}{\sum_j I(f_j = 1)}$$

  where $j = 1, \ldots, J$ denotes possible values in the sample, F_j is the number of individuals in the population with key value j (frequency of j in the population), f_j is the same frequency for the sample and I stands for the cardinality of the selection. Unless the sample size is much smaller than the population size, $P(PU|SU)$ can be dangerously high; in that case, an intruder who locates a unique value in the released sample can be almost certain that there is a single individual in the population with that value, which is very likely to lead to that individual's identification.

  Note that for numerical attributes where the values are almost always quasi-identifiers uniqueness is very high.
- **Record-level risk uniqueness.** They are also known as individual risk measures. Disclosure risk is defined as the probability that a particular sample record is reidentified, i.e. recognized as corresponding to a particular individual in the population. As [41] points out, the main rationale behind this approach is that risk is not homogeneous within a data file. We summarize next the description given in [42] of the record-level risk estimation.

  Assume that there are K possible combinations of key attributes. These combinations induce a partition both in the population and in the released sample. If the frequency of the k-th combination in the population was known to be F_k, then the individual disclosure risk of a record in the sample with the k-th combination of key attributes would be $1/F_k$. Since the population frequencies F_k are generally unknown but the sample frequencies f_k of the combinations are known, the distribution of frequencies F_k given f_k is considered. Under reasonable assumptions, the distribution of $F_k|f_k$ can be modeled as a negative binomial. The per-record risk of disclosure is then measured as the posterior mean of $1/F_k$ with respect to the distribution of $F_k|f_k$.

5.3.4 Reidentification

As explained above, we can define a measure of disclosure risk as the proportion of records that an intruder can reidentify. This approximation is empirical (computational) in the sense that reidentification algorithms are implemented and then used to compare and link the masked data set with the data available to the intruder.

Using the notation in Fig. 5.3 ($A = X' = \rho(X)$: protected set; and $B \subset X$: intruder's set), we define the reidentification risk below using a function $true : B \rightarrow A$ that for each record b (of the intruder) returns the correct record for reidentification, and a function $r : B \rightarrow A$ that models the reidentification algorithm and thus for each record b returns a record in the masked file. We also use a function c such that $c(r(b), true(b))$ is one when both functions return the same record in A.

$$Reid(B, A) = \frac{\sum_{b \in B} c(r(b), true(b))}{|B|}. \tag{5.1}$$

We can consider variations of this definition. For example, when $r(b)$ is a probability distribution on A (and then we can set $c(r(b), true(b))$ as the probability $r(b)[true(b)]$) and when r returns a set instead of a single record (and then we can define $c(r(b), true(b))$ as $1/r(b)$ when $true(b) \in r(b)$ and 0 otherwise).

Another important issue is that reidentification algorithms can return some correct matchings and some incorrect ones, but the intruder may not be able to distinguish among them. This is often the case when data has been masked as e.g. the distance to the nearest and the second nearest may not give enough evidence on whether the nearest record is the true match.

Note that in Eq. 5.1, the set of reidentified records is

$$RI = \{b | r(b) = true(b)\}.$$

Let K be the subset of records in RI such that the intruder is sure that reidentification is correct. Under these premises, we define

$$K.Reid(B, A) = \frac{|K|}{|B|} \tag{5.2}$$

Finally, as for a given dataset both measures can be relevant, we define $KR.Reid$ in terms of the two measures we have introduced.

$$KR.Reid(B, A) = \left(\frac{|K|}{|B|}, \frac{|RI|}{|B|} \right). \tag{5.3}$$

Note that $KR.Reid(X, X') = (v_1, v_2)$ with $v_1 = v_2 = 0.001$ can be considered in some contexts more risky than $v_1 = 0$ and $v_2 = 0.25$. The former example means that at least one record over 1000 has been correctly reidentified and the intruder knows which. In contrast, the second example means that 25% of the records can be reidentified but the intruder cannot distinguish the correct matches from the 75%

of incorrect ones. We will discuss this issue again in Sect. 6.1.1 when we present a specific attack for rank swapping.

We have already seen that reidentification is an approach that is general enough to be applied in different contexts. It can be applied under different assumptions of intruder's knowledge, and under different assumptions on protection procedures. Recall Fig. 5.3 and that we can model a range of attacks by means of representing intruder's information in an intruder's file. We have discussed above the case of an intruder having a database with the same schema as the published data set, but also the case of having a different schema.

This approach has also been used to measure disclosure risk for synthetic data (i.e., data that is constructed using a particular data model—see Sect. 6.3). This is the case of [27] where empirical results are given for three different synthetic data generators. Reference [43] also discusses the use of reidentification algorithms for synthetic data.

As we want to measure risk as accurate as possible, the larger the number of correct reidentifications the better. The maximum number possible will correspond to an upper bound of the risk. We discuss this issue in Sect. 5.3.5 devoted to the analysis of the worst-case scenario.

As reidentification algorithms are the essential component of risk evaluation, we will review different approaches in the remaining part of this chapter. We begin in Sect. 5.4 giving an overview of data matching and integration. Data integration studies how to establish relationships between different elements of the databases for their integration. It includes schema matching (that focuses on establishing relationships between database schema) and data matching (that focuses on the linkage of records). Then, we will focus on particular topics of data matching, giving a special emphasis on record linkage when the intruders have the same schema as the published data. Sections 5.5 and 5.6 describe the two main approaches for record linkage. They are, respectively, probabilistic record linkage and distance-based record linkage. Then, Sect. 5.6.3 presents the use of supervised machine learning to estimate disclosure risk. This is related to the worst-case scenario. We will also review record linkage when intruders have a schema different than the one of the published data (see Sect. 5.7).

The literature presents some other record linkage algorithms, some of which are variations of the ones presented in this book. For example, [44] presents a method based on cluster analysis. This is similar to distance-based record linkage as cluster analysis assigns objects (in this case, records) that are similar (in this case, near) to each other, to the same cluster.

There are algorithms that differ on the type of links allowed between intruder's and published files. Here we focus on algorithms that permit two different records of the intruder b_1 and b_2 to be assigned to the same record a. Others [45–47] force different records in B to be linked to different records in A. When A and B have the same number of records, such algorithms find a one-to-one correspondence between the records of both files. This is known as perfect matching in graph theory. Reference [47] describes the RELAIS record linkage software developed by ISTAT that permits users to consider one-to-one, one-to-many, and many-to-many reidentification problems. Reference [45] presents a detailed analysis of the performance (with respect to number of reidentifications) of this type of algorithms in comparison with the ones not requiring a perfect matching. This latter work is based on the Hungarian algorithm. An $O(n^3)$ combinatorial optimization algorithm that solves the assignment problem.

It is also relevant to mention specific record linkage algorithms which can be used when the transparency principle is used. Thus, they are used to implement transparency attacks. These algorithms take advantage of any information available about the data protection procedure. In other words, protection procedures are analyzed in detail to find flaws that can be used for developing more effective, i.e. with larger matching rates, record linkage algorithms. Attacks tailored for two protection procedures are reported in the literature. Reference [48] was the first specific record linkage approach for microaggregation. More effective algorithms have been proposed in [49,50] (for either univariate and multivariate microaggregation). Reference [30] describes an algorithm for data protection using rank swapping. We will discuss in Sect. 6.1.1 the algorithm for rank swapping. They are related to the transparency property in data privacy mentioned in Sect. 1.3.6.

5.3.5 The Worst-Case Scenario

In any analysis of risk it is important to have an upper bound of the risk. This corresponds to the risk of the worst-case scenario. When we are using reidentification to assess risk, the data available to the intruder and the reidentification algorithm makes a difference on the number of correct links.

The upper bound of the risk can be computed taking into account the following.

- The maximum data an intruder may have is the complete original file, the one that has been protected and later released. In general, given two intruders with files B_1 and B_2, the one with more attributes and more records will be the one able to reidentify more. Nevertheless, more records than the ones in the masked file A do not help neither. Similarly, more attributes than the ones in A do not help. Note that records not in A will not help because they can confuse the reidentification algorithm linking records not present. Similarly, attributes other than the ones in A will provide, at most, the same information already present in A. From this discussion it follows that the worst-case scenario is when intruders' schema is the same as the schema from the published data.

  This discussion applies when the masked file A has been created from a file with the same records and attributes than the ones in A. When the masking uses other records and attributes, these can be useful in the attack if information on the data masking is released (see below).

- The maximum information on the masking procedure an intruder may have is the masking method and its parameters. The transparency principle states that when data is released we need to inform about all the processes applied to the data (see Sect. 1.3.6). In our case, this includes the data masking as well as its parameters. If this information is available, the intruder can use this information to attack the database and can increase the number of correct links. Deterministic methods can be easily attacked with this information. Specific record linkage methods are the ones developed taking into account this information. This is similar to the scenarios considered in cryptography where intruders know the algorithms but not the keys. Here, the algorithms are known but not e.g. the random numbers used in the masking.

- The maximum number of reidentifications is obtained with the best reidentification algorithm using the best parameters.

  - There are different approaches for reidentification. We have mentioned probabilistic and distance-based record linkage. The worst-case scenario corresponds to the case with the largest number of reidentifications. Therefore, among different algorithms, we presume that the intruder will use the one that performs better (i.e., the one that leads to the maximum number of reidentifications).
  - Most reidentification algorithms are parametric. In this case, different parameterizations may lead to different number of reidentifications. The worst-case scenario, which correspond to the upper bound of the risk, corresponds to the parameters that lead to the largest number of reidentifications. We can use machine learning and optimization algorithms to find these successful parameterizations for given datasets. This is discussed in Sect. 5.6.3.

As a summary, for $X' = \rho_p(X)$ where ρ is the masking method applied to the data set X and p its parameters, the worst-case scenario corresponds to compute $KR.Reid(X, X')$ (from Eq. 5.3) where the reidentification algorithm r takes advantage of both ρ and p.

5.4 Matching and Integration: A Database Based Approach

We have claimed that reidentification, and, in general, matching, is a flexible way to model and assess disclosure risk. Database integration supplies us with a set of tools that can be used for this purpose. In this chapter we focus on tools related to this area.

Nevertheless, as there is a large number of information sources from which we can extract information useful to attack a database, there are still other tools that can be used. As an example of the diversity of sources, recall the Netflix case [51] where the data in the Netflix database was linked to data from Internet Movie Database (IMDb) users. Tools useful in this context include natural language processing algorithms and ontology matching software. On the use of ontologies for reidentification see e.g. [52].

5.4.1 Heterogenous Distributed Databases

It is well known that available data are not centralized but highly distributed. It is usually the case that records corresponding to a particular person are spread within several subsidiary companies or departments of the same organization (e.g. sales department, customer department and suppliers department).

In order to take full advantage of available data in distributed databases, software systems have been defined to find and link records that while belonging to different databases they correspond to the same individual. Although it could be possible that the different databases have the same software and the same schema (i.e., we have an homogeneous database) the most common and problematic case is that they have different software and different schema. In this case, the *set of different databases* and the corresponding integration system is known in the literature by *heterogeneous database*. Reference [53] defines heterogeneous databases as follows:

> A Heterogeneous database consists of a set of interconnected, autonomous component databases. The components communicate in order to exchange information and answer queries. Objects in one component database may differ greatly from objects in other component databases, making it difficult to assimilate their semantics into the overall heterogeneous databases.

Difficulties in the integration in a heterogeneous database are due to the fact that different databases contain partially overlapping data and this overlapping information is usually non-uniformly standardized and subject to all kind of errors and accuracies. Robust integration methods are needed to avoid that the result of a query is incorrect or inconsistent with previously reported results.

Data integration is the field that studies how data from different databases are combined in a single one. This is an active area of research, and there exist nowadays several commercial solutions for this purpose. Some of the existing solutions are for a virtual data integration. In this case, there is a wrapper based on a global schema but there is no real integration of the databases. Some others materialize this integration.

5.4.2 Data Integration

There exist several technologies for data integration. Let us discuss first some of the problems we can encounter in data integration. We consider four different scenarios and the most typical problems. This follows [54]. The scenarios are classified in two dimensions. As some of the problems are specific to data integration (i.e., to multiple databases) and others also appear in scenarios of a single database (e.g., the problem of finding an individual in a database or the problem of database deduplication) one dimension is the number of sources. The other dimension is the problem level.

- **Number of sources**. That is, the number of data suppliers considered in the process. We consider single source and multi-source (multi-database) problems.

- **Problem level**. Inconsistencies and integration difficulties can appear in both instance (or record) level and schema level. Schema level problems are the ones that can be solved with an appropriate schema and the corresponding integrity constraints. Naturally, instance level problems are the ones strictly related to the data represented in the records of the database.

We review below the four scenarios that can be obtained according to these two dimensions.

- **Single source at instance level (SSIL)**. In this scenario we have data integration in deduplication (removal of duplicate records in a database). Typical problems are misspellings, missing values but also contradictory records (e.g., different addresses for the same person).
- **Single source at schema level (SSSL)**. Here problems typically arise because the database schema is not strict enough and some integrity constraints are not checked. This is the case of illegal values (e.g. a record with Name = John, Sex = Male, Pregnant=true) or inconsistent values (e.g. current date $\neq$ birthday + age).
- **Multi-source at instance level (MSIL)**. Problems specific to this scenario include differences on representation formalisms. For example, different databases may use different recodifications, different scales or different granularity. For different recodification, let us consider one database using the 10-th revision of the International Classification of Diseases [55] to codify illnesses while the second one uses a full text description of the illness. For different scales consider Celsius versus Fahrenheit scales for temperature. For different granularity consider one database where fever is represented using real numbers (from data recorded in a hospital) and another using linguistic terms (e.g., no-fever, a little) and numbers in a mixed way because data was collected in interviews with the patients.
- **Multi-source at schema level (MSSL)**. Problems correspond to naming and structural conflicts. With respect to names we may have the same name for different objects (homonyms) and different names for the same object (synonyms). More complex structural problems arise when one attribute in one database (e.g. address) correspond to several attributes in the other (street, zip code, city).

In Table 5.2, we report some of the problems typically found in these four scenarios (see [54]).

The distinction between schema and data matching problems is not always clear. Some problems can be considered from the two perspectives. For example, the problem with the temperature scale and the divergence of illness codification can be seen as a problem at instance level but also as a schema level problem if we consider that the schema should include information about scales and codification.

In the remaining part of this section we further elaborate into the problems related to data matching. Now we focus on the simpler problem of considering two databases and a single attribute in each database. Although we can consider in this case that, again, some of the problems are due to discrepancies on the schema, we consider

Table 5.2 Single and multi-source (SS/MS) problems at instance and schema level (IL/SL) based on [54]

	Problem	Example
SSIL	Missing values	Null values
	Misspellings	
	Misfielded values	
	Unknown abbreviations	
	Multiple variable: non-consistent use	Different order: name/first name "Joan John" versus "John, Joan"
	Contradictory records	Same name/different address: "Joan John, Alpha street" "Joan John, Beta street"
	Wrong or outdated data	
SSSL	Illegal values	Data outside rang, negative age
	Violation of attribute dependencies	Price $\neq$ net price + taxes City and zip code not consistent
	Uniqueness violation	Passport not unique
	Referential integrity violation	Reference to an unknown record
MSIL	Different scales	Temperatures: Celsius versus Fahrenheit
	Different codifications	True/false 1/0 t/f t/nil
	SSIL problems	
MSSL	Name conflicts	Synonyms: customer/client, sex/gender
	Structural conflicts	(Address) versus (num. str. zip city): "203, Alfa street 08201 Sabadell" "Alfa st." "2003" "08201" "Sabadell"
	SSSL problems	

them as divergences on the attributes. We classify the situations according to two dimensions. One is about the coincidence on variables and the other one is about the coincidence on terminology. A graphical interpretation of this classification is given in Fig. 5.4.

- **Coincidence on variables**. Variables describing data can be the same in both databases or can be (partially) different.
- **Coincidence on terminology**. Terminology in this framework corresponds to the domain on the variables. That is, terminology is the set of terms used to evaluate the individuals. As before, two cases are considered: equal and different terminology. By the way, in some situations it would be appropriate to distinguish a third case corresponding to small differences between variables domain (*small* from a semantics point of view). This latter case includes the case of terms that although different when considered from its syntactical definition (e.g. the terms "low" and "small" are different) they refer to the same concept or idea. Similarly,

Fig. 5.4 Variable-related
dimensions in data
integration problems

Terminology (variable domains)

		SAME	DIFFERENT
Variables	SAME	consensus	correspondence
	DIFFERENT	conflict	contrast

we classify in this group the case of two terms that are subject to errors (for example, surnames with misspelling: *Green* vs. *Grene*).

The combination of both dimensions leads to four different cases (see Fig. 5.4). Following [56], the following terms are used for describing these cases.

- **Consensus**. The databases use both variables and terminology in the same way. In this case, records can be moved from one database to the other with no changes.
- **Correspondence**. The database use the same variables but the terminology for describing these variables are different. For example both databases contain a variable named *location* but while in one database it corresponds to *town* in the other refers to *counties*. In this case, a mapping between both terminologies are required. The example given above about a variable *temperature* expressed either in Celsius or in Fahrenheit also belongs to this case.
- **Contrast**. This case corresponds to using the same terminology for different variables. It is important to note that different variable names in different databases can either correspond to a unique underlying variable or to two different variables (i.e., equal or different from a semantics point of view). An example of the first situation is when we consider a variable *city* and another *location* both containing city names. An example of the latter situation is when we consider the variables *work/city* and *city* both containing city names but one referring to a work-place and the other to household place.
- **Conflict**. The last case is when databases differ in both terminology and variable names.

This classification is based on [56] (for knowledge elicitation methods from groups) and [57]. Reference [56] considers the problem of integrating grid data [58] from two experts.

In order to solve the problems described in this section, data integration considers three tasks (see [59]).

- **Schema matching**. This is to find correspondences between schemas of two or more databases.
- **Data matching**. Presuming that different databases use the same schema, this is to find which objects in one database correspond to which objects in another database.
- **Data fusion**. This is to consolidate in a single database the data that is known to correspond to the same object.

From the point of view of disclosure risk assessment only schema and data matching are of interest. Schema and data matching are discussed below in Sects. 5.4.3 and 5.4.4, respectively.

There are several ways to fuse and consolidate data. For example, we can use the mean of the values or select the most recent one. A discussion on aggregation functions and data fusion methods from the point of view of their mathematical properties can be found in [60].

We will see that record linkage is used in one of the steps of data matching. In data privacy, record linkage is mainly used when we have *consensus* between the two databases, and when we have *correspondence* with small variations in the terminology (e.g. due to transcription errors), or with a clear correspondence between the set of terms. Knowledge-based record linkage (as in [52]) based on ontologies can be used in this latter case. Ontologies are also used in schema matching. We will review some of the existing approaches in the next section. Contrast and conflict is not considered.

The use of semantics in data privacy to attack a database links data integration, schema and data matching with ontology matching. We will not address this topic further. For a recent overview of ontology matching, see e.g. [61,62].

5.4.3 Schema Matching

As the name indicates, schema matching is about establishing correspondences between the schema of two or more databases. The goal is to match variables that represent the same information.

One of the difficulties of this task is that databases may be defined in terms of a large number of variables, and this causes the matching process to work with a combinatorial explosion. Another difficulty is that not only $1 : 1$ relationships are possible (one variable in one database with another one in another database), but also $1 : n$ or $m : n$ relationships. That is, it may be possible that m variables in one file correspond to n variables in another file. Name and address are typical examples of such relationships. We may have one or more fields for title, names and surnames. Similarly, address can be splitted into different components in the databases.

Attribute (variable) correspondence identification methods focus on finding correspondences between pairs of variables.

In the literature we find methods that are solely based on variables names and data types while others use some kind of structural information [63,64] or characteristics of data instances [65]. Other methods combine the information in the database with some other information or, following Artificial Intelligence terminology, background knowledge. This corresponds to knowledge about the variables or terms that can be represented in ontologies or dictionaries (as e.g. Wordnet [66]). References [63,65] describe one of such systems. Hybrid approaches have also been considered. They combine several identification methods following the machine learning approach of combining several classifiers. One of such systems is COMA [64] that provides an extensible library off variable-matching algorithms and supports different ways for combining matching results.

Structure-level matching methods match combination of elements. Reference [67] gives a broad account of such methods.

Operators for data transformation can help schema matching. For example, [68] uses user-defined functions supported in SQL for some data transformations. In general, such operations permit us to define new views from an initial table so that the schema matching procedure can be found more easily. This is the case, for example, of decomposing a *raw* address field in a file, say A, into the following variables: Address, City, Zip code, State and Country. In this way, the Zip code in a file B can be easily matched with the corresponding Zip code in file A.

Data analysis tools are also helpful in schema matching. They can be used to increase our knowledge about the data and extend meta-data description. On the one hand we have data profiling tools, which extract basic information about data types (type, length, typical patterns, ...) and basic statistics about the data (frequency or variance, value range). For example, the method described in [31] permits to characterize the probability distributions of the variables and then link variables in the two files with the same distributions. On the other hand we have tools for model building. They will find relationships between the variables that can be latter used for matching. For example, regression models and decision trees can be used to establish relationships between variables. Such analysis can help the user to find incorrect records. For example, records not compliant with high frequency models might contain errors.

Machine learning has been used as a data analysis tool for finding typical patterns in complex string variables. Reference [69] describes a tool for extracting structured elements like *Street name*, *Company* and *City name* from an address record occurring as a free text string. The method is based on Hidden Markov Models (HMM). Reference [70] presents a similar system also based on the hidden Markov model. Methods for automaton and grammar determination can also be used in similar contexts.

In some situations, it is not even known if databases share variables. SEMINT [71], a tool for identifying relationships between variables in different database schemas, was developed to deal with this problem. It is based on neural networks and uses both schema information and data contents for building such relationships.

Note that all these tools can be used by an intruder whose database has different schema than the one published.

5.4.4 Data Matching

Data matching focuses on establishing relationships between the records. The goal is to identify the records that belong to the same individual but that are in different databases. This problem is sometimes known (e.g., [54]) as the object identity problem and the merge/purge problem.

Given two databases, the process of matching them can be described in terms of the five steps (see [59] for more details) described below. Note that there is some overlapping between the steps here and the problems considered in schema matching.

- **Data preprocessing**. In this step data files are transformed so that all attributes have the same structure and the data have the same format. In data fusion, this is said to make data commensurate. That is, data should refer to the same point in time and refer to the same position in space. The same should be done here to make data comparable. Major steps in data preprocessing (see [59], Section 2.2) are: (i) remove unwanted characters and words (stop words), (ii) expand abbreviations and correct misspellings, (iii) segment attributes into well defined and consistent output attributes, (iv) verify correctness of attributes values (e.g., that ages are always positive). Note that (iii) is related to schema matching.
- **Indexing**. In most data matching problems it is unfeasible to compare all pairs of records in order to know which pairs correspond to the same individual. Note that the number of comparisons is the product of the number of records in the two databases. In order to reduce the number of comparisons, only some pairs are compared. Indexing is about the determination of which are the pairs interesting to be compared.
- **Record pair comparison**. This consists in the calculation of a value for each pair of records of interest. The comparison can be either a vector of Boolean values (stating whether each pair of attributes coincide or not) or a vector of *similarities* (stating in a quantitative way how similar are the values of the corresponding attributes).
- **Classification**. Using the comparison we need to establish whether the two records in the pair correspond to the same object or they correspond to different objects.
- **Evaluation step**. The result of the data matching system is analyzed and evaluated to know its performance.

We discuss the steps related to data privacy in more detail below. Preprocessing in Sect. 5.4.5, indexing in Sect. 5.4.6, record pair comparison in Sect. 5.4.7 and classification in Sect. 5.4.8.

We do not discuss here the evaluation step of data matching systems. Instead, we will consider in the sections devoted to record linkage (Sects. 5.5.3 and 5.6.3) how to tune the parameters of the system.

5.4.5 Preprocessing

Assuming that there is no discrepancy between variables, data need to have the same format. Different forms of the same name (e.g., Robert, Bob), company names (e.g., Limited, Ltd., LTD) and addresses (e.g. street, st.) are transformed into a single form. This step can be seen as a final standardization step prior to the comparison. It consists of:

1. Parsing variables to build a uniform structure.
2. Detecting relevant keywords to help in the process of recognizing the components that form the values of a variable.
3. Replacing all the (common) forms of a word by a single one (for example, expand abbreviations).

The goal of parsing is to ensure that, when the value of the variable consists of several elements, these always appear in the same order. For example "Robert Green, PhD", "Dr. Bob Green" and "Green, Robert" are translated into "PhD Robert Green", "Dr. Bob Green" and "Robert Green", respectively, following a *title + name + surname* structure.

The detection of special keywords can help in this process. For example, detection of "Ms", "PhD" or "Dr" is usually an indication of the presence of a personal name and "Ltd" indicates the presence of a company name. Detection would trigger specific parsing routines when appropriate.

The third standardization procedure replaces variants of values by a standard form. Depending on the meaning and the values of the variable, this procedure can either be applied to the whole variable value (e.g., to the string used to represent the name) or to components of the variable value. This latter case occurs when the variable corresponds to personal names and they include for example title and middle letters, or when the variable is an address with street names, numbers or a P. O. box.

The substitutions required by standardization can be efficiently implemented by building a database with lists of words and their corresponding standard form so that the forms that appear in the files can be *replaced* by the standard ones. It is important to note that this *standard* form does not need to be a "dictionary" form (the root of a word or any not-shortened version of the name) but only an abstract identifier. This abstract identifier can be useful when a single spelling can have different origins (e.g. Bobbie might refer to Robert but also to Roberta).

For details on standardization see [72]. Examples of name and address parsing are provided there.

5.4.6   Indexing and Blocking

Let **A** and **B** be the files to be linked. When these files contain a huge number of records, considering all possible pairs is rather costly. Note that we need to consider $|\mathbf{A}| \cdot |\mathbf{B}|$ pairs. Among these pairs, only $\min(|\mathbf{A}|, |\mathbf{B}|)$ pairs can be effectively linked when each individual appears at most once in each file. Blocking methods are defined to eliminate most of the unsuccessful comparisons. They define blocks of records and then comparisons between records are restricted to records in the same block. There are different approaches to blocking. Based on [73] we classify them into three categories.

- **Blocking variables**. Variables are sometimes used for this purpose. They are selected by the user among the most error-free ones present in both files (those variables most likely to maintain their values across files). Only records with equal values for all blocking variables are compared. To implement this approach, files are ordered according to blocking variables.

  Typical examples of blocking variables are the gender, ZIP code and other geographical variables. When string variables are used, a good alternative is to use the first letter or a particular coding so that all the symbols with a similar sound are mapped onto the same block (for example, the SOUNDEX codification—see Definition 5.5 and Sect. 5.4.7). This is to reduce the possibility that records that should be compared are in different blocks.
- **Cluster-based blocking**. In this approach, records are clustered and comparisons are only applied within records in the cluster. In this case, blocks are built not on the basis of a single variable but a set of them, expecting that errors will not make records that should be linked together differ too much.

  The use of clustering methods that require a large number of comparisons (between pairs of record pairs) is discouraged. This will not reduce the cost of record linkage.
- **Locality-sensitive hashing**. Hashing functions[3] are used so that similar records are mapped into the same blocks. Functions are selected to maximize the number of linked records that are mapped into the same block. Reference [75] presents a survey of some indexing techniques, and compares them experimentally. Reference [73] also focuses on this type of blocking method. The semantic blocking proposed in [76] also falls in this area.

As blocking methods are not always correct, some of the linked pairs are not detected. I.e., we will have *missed matches*. An unsuitable selection of the blocking method can result in a large number of missed matches. An approach to reduce the negative effects is to apply several times the record linkage method using in each iteration a different blocking method (e.g., a different blocking variable). Naturally, this process increases the complexity of the data matching process. Blocking methods correspond to a compromise between a high-cost detailed analysis of all possible

---

[3] A description of hash functions, very common in data structures, can be found e.g. in [74].

pairs with few missed matches and a low-cost analysis of only a few pairs with
more missed matches. To help in the process of defining the blocking method, [77]
introduced a machine learning approach.

5.4.7 Record Pair Comparison: Distances and Similarities

The comparison between any pair of records is rarely based only on a binary coincidence of the values in the two records. In real-world databases, data includes errors.
This is similar in data privacy although in this case masking methods introduce noise
in the data on purpose to decrease disclosure risk. Because of that, comparison is usually expressed in terms of a distance between pairs of records. More specifically, the
distance between pairs of records is usually expressed in terms of specific distances
for each variable.

Due to this, we can consider distances at record level and distances at variable
level. The former will be defined in terms of the latter. We discuss in this section the
basics of distances between records, focusing on the distances between variables.
Other distances at record level will be discussed in Sect. 5.6.

One of the most usual definitions is to consider that all variables have equal weight
and define the distance between two records as the addition of the distance between
values. This definition follows.

Definition 5.2 Let a and b two records defined on variables $\{V_1, V_2, \cdots V_n\}$. Assuming equal weight for all variables, we define $d(a, b)$ as follows.

$$d(a, b) = \sqrt{\sum_{i=1}^{n} d_{V_i}^2 (V_i^A(a), V_i^B(b))}$$

where $d_{V_i}^2 (a_i, b_i)$ is the square of a distance on variable V_i.

Recall that formally given a set X and a real function on $X \times X$, d is a metric on
X if and only if for all $a, b, c \in X$ d is positive ($d(a, b) \geq 0$ with $d(a, b) = 0$ only
if $a = b$), d is symmetric ($d(a, b) = d(b, a)$) and d satisfies the triangle inequality
($d(a, b) \leq d(a, c) + d(c, b)$). Then, we call the number $d(a, b)$ the distance between
a and b with respect to the metric d. See e.g. [78]. Note also that we can define
similarities from distances as follows when the distance is bounded by 1: *similarity*
= *f(distance)* (with $f(0) = 1$ and decreasing) as e.g. *similarity = 1 – distance*. Note
that triangle inequality will not hold for a distance defined from a similarity if an
equivalent property does not hold for the similarity.

Distances at variable level depend on the type of variables. We review here some
distances used in the literature of data privacy for numerical, ordinal, and nominal
attributes. See e.g., [18,27]. We will discuss later distances for strings. Alternatively,

if V_i are partitions we can use the expressions discussed in Sect. 2.3.1. Note that this review is by no means exhaustive. For any type of data the literature on distances is large.

Definition 5.3 The following distances have been used extensively for numerical, ordinal, and nominal variables.

1. For a numerical variable V, it is usual to use the expression above with $d_V^2(c, d) = (c - d)^2$. In order to avoid scaling problems, it is convenient to standardize the variables. This results into the Euclidean distance.
2. For an ordinal variable V, let $\leq_V$ be the total order operator over the range of V. Then, the distance between categories c and c' can be defined as the number of categories between the minimum and the maximum of c and c' divided by the cardinality of the range:

$$d_V(c, c') = \frac{|c'' : min(c, c') \leq_V c'' \leq_V max(c, c')|}{|D(V)|}$$

3. For a nominal variable V, the only permitted operation is comparison for equality. This leads to the following definition:

$$d_V(c, c') = \begin{cases} 0 \text{ if } c = c' \\ 1 \text{ if } c \neq c' \end{cases}$$

where c and c' correspond to categories for variable V.

In the case of nominal variables for which a hierarchical structure has been defined, [79] used the following distance which is based on the closest common generalization of two given categories. The closest common generalization of two categories c_1 and c_2 corresponds to the nearest category that has both c_1 and c_2 as descendants.

Definition 5.4 Let H be a hierarchical structure representing the categories of a variable V. Let $dpt(c)$ be the depth of the category c in the hierarchical structure, and let $CCG(c_1, c_2)$ the closest common generalization of categories c_1 and c_2. Then, the distance between c_1 and c_2 is defined as

$$distance(c_1, c_2) = dpt(c_1) + dpt(c_2) - 2dpt(CCG(c_1, c_2)).$$

There is large number of distances defined for strings (see e.g. [80,81] for reviews). We start reviewing SOUNDEX, which transforms a string into a code that tends to bring together all variants of the same name. Therefore, the application of this method for string comparison leads to a Boolean comparison (strings are either encoded in the same or in a different way). A description of this method, originally developed by Odell and Russell in [82,83], can be found in [84]. The SOUNDEX code of any string is a sequence of four characters. For example, both strings "Smith" and "Smythe" are encoded as "S530". Then, comparison between strings is achieved by means of comparison of sequences. This coding has been used to deal with surnames. Jaro [85] recommends its use as a blocking variable in the indexing step.

> To maximize the chance that similarly spelled surnames reside in the same block, the SOUNDEX system can be used to code the names, and the SOUNDEX code can be used as a blocking variable. There are better encoding schemes than SOUNDEX, but SOUNDEX with relatively few states and poor discrimination helps ensure that misspelled names receive the same code (Jaro 1989, p. 418 [85])

Reference [85] only recommends it for blocking variables, in other cases it does not recommend its application because nonphonetic errors leads to different codes. Reference [86] states that the coding is effective except when the names are of Oriental origin.

Definition 5.5 Given a string, the SOUNDEX method encodes it into a sequence of one character and three digits as follows.

1. The first letter of the string is selected and used as the first character of the codification.
2. Vowels A, E, I, O, U and letter Y are not encoded. Letters W and H are also ignored.
3. All the other letters are encoded as follows:

$$B, F, P, V \qquad\qquad \text{encoded as 1}$$
$$C, G, J, K, Q, S, X, Z \text{ encoded as 2}$$
$$D, T \qquad\qquad\qquad \text{encoded as 3}$$
$$L \qquad\qquad\qquad\quad \text{encoded as 4}$$
$$M, N \qquad\qquad\qquad \text{encoded as 5}$$
$$R \qquad\qquad\qquad\quad \text{encoded as 6}$$

4. When the coding results into two or more adjacent codes with the same value only one code is kept. The others are removed. E.g., "S22" is reduced to "S2" and "S221" to "S21".
5. All strings are encoded into a string with the following structure: Letter, digit, digit, digit. Additional elements are truncated and in case the string is too short, additional "0" are appended.

Table 5.3 displays some examples taken from Knuth [84]. Examples of pairs of surnames that do not lead to the same codification include (*Rogers, Rodgers*) and (*Tchebysheff, Chebyshev*).

Table 5.3 SOUNDEX codification

Surnames		Coding
Euler	Ellery	E460
Gauss	Ghosh	G200
Hilbert	Heilbronn	H416
Knuth	Kant	K530
Lloyd	Ladd	L300
Lukasiewicz	Lissajous	L222

There exist other methods that proceed in a way similar to SOUNDEX by transforming a large number of strings into a single codification. These methods are classified in [81] as hashing techniques. For example, Blair [87] builds the so-called r-letter abbreviations. This procedure transforms all strings s to r-letter strings removing $length(s) - r$ irrelevant characters. In this method, relevance of a character is computed in terms of relevance of letters (e.g., "A" has relevance 5 and "B" relevance 1) and relevance of position (e.g., relevance of second position is larger than relevance of first position). Some example codings for 4-letter abbreviations are: *Euler* and *Ellery* are transformed to ELER and *Tchebysheff* and *Chebyshev* are transformed to ESHE. *Rogers* is translated to OERS and *Rodgers* can either be translated to OERS or GERS (letters "O" and "G" in *Rodgers* have the same importance but only one of them can be deleted).

Bigrams have been used to compare strings and compute a similarity measure. A bigram is defined as a pair of consecutive letters in a string. Therefore, the word bigram contains the following bigrams: bi, ig, gr, ra, am. The value of the function *simB* applied to two strings *s1* and *s2* is a value in the [0, 1] interval corresponding to the number of bigrams in common divided by the mean value of bigrams in both strings:

$$simB(s1, s2) = \frac{|bigrams(s1) \cap bigrams(s2)|}{(|bigrams(s1)| + |bigrams(s2)|)/2}$$

where $bigrams(s)$ corresponds to the bigrams in string s.

Naturally, this function defines a similarity function and is equal to 1 when both strings are equal.

As said above, bigrams correspond to two consecutive characters. In fact, the literature also considers the general structure of n-grams (n consecutive characters in a string—a substring of length n). Similarity measures have been considered for n-grams with $n > 2$.

Another approach is the Jaro algorithm, introduced in [88].

Definition 5.6 The similarity between two strings $s1$ and $s2$ according to the Jaro algorithm is defined as follows.

1. Compute the length of the strings $s1$ and $s2$:

   - $strLen1 = length(s1)$
   - $strLen2 = length(s2)$

2. Find the number of common characters. These characters are the ones that appear in both strings at a distance that is at most $minLen/2$ where $minLen = \min(strLen1, strLen2)$:

   $$common = \{c | c \in chars(s1) \cap chars(s2) \text{ and } pos(s1) - pos(s2) \leq minLen/2\}$$

3. Find the number of transpositions among the common characters. A transposition happens whenever a common character from one string does not appear in the same position as the corresponding character from the other string. Let $trans$ be the number of transpositions.

Then, the Jaro similarity is defined as follows:

$$jaro(s1, s2) = \frac{1}{3} \left(\frac{common}{strLen1} + \frac{common}{strLen2} + \frac{1}{2} \frac{trans}{common} \right)$$

McLaughlin, Winkler and Lynch have studied this similarity and defined some enhancements which, when combined with record linkage, improve the performance of the latter. The enhancements and results are given in [89] (see [90] about the code for this distance).

Another approach for computing similarities between strings is based on dynamic programming. Algorithm 16 describes how to compute the Levenshtein distance [91] between two strings. This distance is defined on any pair of strings (not necessarily of the same length) and gives the same weight (assumed to be 1 in the algorithm below) to insertions, deletions and substitutions.

Improvements of this method exist so that the computation time and the working space requirement are reduced. See [81] for details.

Reference [92] presents a comparison of several efficient string distances in the particular context of record matching. Reference [93] proposed recently another distance which satisfies all axioms of a distance and is also efficient.

5.4.8 Classification of Record Pairs

Record linkage algorithms are used to classify pairs of records. There are two main approaches.

Algorithm 16: Levenshtein distance.

Data: $s1$: string; $s2$: string
Result: Levenshtein distance between $s1$ and $s2$

1 **for** $i = 0$ **to** m **do**
2      $d[i,0] = i$;

3 **for** $i = 0$ **to** n **do**
4      $d[0,i] = 0$;

5 **for** $j = 1$ **to** n **do**
6   **for** $i = 0$ **to** m **do**
7   **if** $s1[i] == s2[j]$ **then**
8              $d[i, j] = d[i - 1, j - 1]$;
9   **else**
10              $d[i, j] = 1 + \min(d[i - 1, j], d[i, j - 1], d[i - 1, j - 1])$;

11 **return** $d[m, n]$;

- **Distance-based record linkage methods**. They assign each record in one file to the nearest record in the other file. They are simple to implement and to operate. The main difficulty consists of establishing appropriate distances for the variables under consideration. The advantage of distance-based record linkage is that it allows inclusion of subjective information (individuals or variables can be weighted) in the reidentification process.
- **Probabilistic record linkage methods**. They are less simple to implement. Assignment is done according to a probabilistic model. However, they do not assume rescaling or weighting of variables and require the user to provide only two probabilities as input. Probability of having false positives and false negatives.

Other record linkage methods include cluster-based and rank-based record linkage. The cluster-based approach (see [44]) is based on merging the two files in a single file and then building a large number of clusters (ideally, equal to the number of people) expecting records from the same individuals to be located in the same cluster. Rank-based record linkage was recently introduced in [94] and consists of computing the distance on ranked data instead on the original values. That is, original values are ordered and each value is replaced by its position in the order. Distance-based and probabilistic record linkage can then be applied to these ranks.

Probabilistic record linkage is described in Sect. 5.5 and distance-based record linkage in Sect. 5.6. The description of these two methods is based on [95]. These two methods are applicable when the two files have the same set of variables. In Sect. 5.7 we consider the case of files not sharing variables. Section 5.9 compares the different approaches according to our experimental results.

In order to compare in what extent the number of correct reidentifications an intruder obtains can be due to randomness, we give below the following result.

Proposition 5.1 *[57] Let A and B be two files corresponding to the same set of n individuals. The probability of finding at random a permutation with exactly r records in the correct position is*

$$\frac{\sum_{v=0}^{k} \frac{(-1)^k}{v!}}{(n-k)!} \tag{5.4}$$

where $k := n - r$.

In most real problems, the probability of finding a few reidentifications by chance is negligible. Reference [57] gives the probabilities in the case of 35 objects, and finding more than or equal to 3 correct links is less than 0.1.

5.5 Probabilistic Record Linkage

The goal of probabilistic record linkage is to establish whether pairs of records $(a, b) \in \mathbf{A} \times \mathbf{B}$ either belong to the set $\mathbf{M}$ or to the set $\mathbf{U}$, where $\mathbf{M}$ is the set of matches (i.e., correct links between records in $\mathbf{A}$ and $\mathbf{B}$) and $\mathbf{U}$ is the set of unmatches (i.e., incorrect links). That is, whether both records a and b correspond to the same individual or to different individuals. Decision rules classify pairs as linked or non-linked by placing them, respectively in $\mathbf{M}$ and $\mathbf{U}$). Some decision rules in probabilistic record linkage consider an additional classification alternative: clerical pairs. They are the ones that cannot be automatically classified neither in $\mathbf{M}$ nor in $\mathbf{U}$. Classification of clerical pairs is done manually. As a summary, the following classes are considered by decision rules: $\mathbf{DR} = \{\mathbf{LP}, \mathbf{CP}, \mathbf{NP}\}$.

1. **LP**: Set of linked pairs
2. **CP**: Set of clerical pairs
3. **NP**: Set of non-linked pairs

In probabilistic record linkage, a basic assumption is that files share a set of variables. Taking this into account, decision rules rl are defined as mappings from the comparison space (the space of all comparisons Γ) into probability distributions over $\mathbf{DR}$. If $\gamma \in \Gamma$ then $rl(\gamma) = (\alpha_1, \alpha_2, \alpha_3)$ where $\alpha_1, \alpha_2, \alpha_3$ are, respectively, the membership probabilities for $\{\mathbf{LP}, \mathbf{CP}, \mathbf{NP}\}$. Naturally, $\alpha_1 + \alpha_2 + \alpha_3 = 1$ and $\alpha_i \geq 0$.

Example 5.2 (From [95]) Let us consider the files $\mathbf{A}$ and $\mathbf{B}$ in Table 5.4. Both files contain 8 records and 3 variables (Name, Surname and Age). For the sake of understandability, the files are defined so that records in the same row correspond to matched pairs and records in different rows correspond to unmatched pairs. The goal of record linkage in this example is to classify all possible pairs so that pairs with

Table 5.4 Files **A** and **B** used in Example 5.2

Name^A	Surname^A	Age^A	Name^B	Surname^B	Age^B
Joana	Casanoves	19	Joana	Casanovas	19
Petra	Joan	17	Petra	Joan	17
J.M.	Casanovas	35	J. Manela	Casanovas	35
Johanna	Garcia	53	Johanna	Garcia	53
Ricardo	Garcia	14	Ricard	Garcia	14
Petra	Garcia	18	Petra	Garcia	82
Johanna	Garcia	18	Johanna	Garcia	18
Ricard	Tanaka	14	Ricard	Tanaka	18

both records in the same row are classified as linked pairs and all the other pairs are classified as non-linked pairs.

To do so, we consider all pairs $(a, b) \in \mathbf{A} \times \mathbf{B}$. These pairs and the corresponding $\gamma(a, b)$ are displayed in Table 5.5. Note that $\gamma(a, b)$ is a coincidence vector in Γ. In this example, $\Gamma = \{\gamma^1 = 000, \gamma^2 = 001, \gamma^3 = 010, \gamma^4 = 011, \gamma^5 = 100, \gamma^6 = 101, \gamma^7 = 110, \gamma^8 = 111\}$. Note that the number of different coincidence vectors (8) is much less than the number of pairs in $\mathbf{A} \times \mathbf{B}$ (64). Yet, in probabilistic record linkage, the classification of any pair (a, b) in Table 5.5 is solely based on its corresponding coincidence vector $\gamma(a, b)$.

Decision rules are defined for $\gamma \in \Gamma$ and they are based on the following expression:

$$\frac{P(\gamma = \gamma(a, b)|(a, b) \in \mathbf{M})}{P(\gamma = \gamma(a, b)|(a, b) \in \mathbf{U})}. \tag{5.5}$$

Definition 5.7 Let (a, b) be a pair of records in $\mathbf{A} \times \mathbf{B}$ and let (lt, ut) be two thresholds (lower and upper) in $\mathbb{R}$ such that $lt < ut$. Then the FS (for Fellegi and Sunter [96]) decision rule is:

1. If $R_p(a, b) \geq ut$ then (a, b) is a Linked Pair (**LP**)
2. If $R_p(a, b) \leq lt$ then (a, b) is a Non linked Pair (**NP**)
3. If $lt < R_p(a, b) < ut$ then (a, b) is a Clerical Pair (**CP**)

where the index $R_p(a, b)$ is defined in terms of the vector of coincidences $\gamma(a, b)$ as follows:

$$R_p(a, b) = R(\gamma(a, b)) = \log\left(\frac{P(\gamma(a, b) = \gamma(a', b')|(a', b') \in \mathbf{M})}{P(\gamma(a, b) = \gamma(a', b')|(a', b') \in \mathbf{U})}\right). \tag{5.6}$$

Table 5.5 Product space $\mathbf{A} \times \mathbf{B}$ and corresponding Γ vectors.

Name^A	Surname^A	Age^A	Name^B	Surname^B	Age^B	$\gamma(a,b)$	$\gamma(a,b)$
Joana	Casanoves	19	Joana	Casanovas	19	101	γ^6
Joana	Casanoves	19	Petra	Joan	18	000	γ^1
Joana	Casanoves	19	J. Manela	Casanovas	35	010	γ^3
Joana	Casanoves	19	Johanna	Garcia	53	000	γ^1
Joana	Casanoves	19	Ricard	Garcia	14	000	γ^1
Joana	Casanoves	19	Petra	Garcia	82	000	γ^1
Joana	Casanoves	19	Johanna	Garcia	18	000	γ^1
Joana	Casanoves	19	Ricard	Tanaka	18	000	γ^1
Petra	Joan	17	Joana	Casanovas	19	000	γ^1
Petra	Joan	17	Petra	Joan	18	110	γ^7
Petra	Joan	17	J. Manela	Casanovas	35	000	γ^1
Petra	Joan	17	Johanna	Garcia	53	000	γ^1
Petra	Joan	17	Ricard	Garcia	14	000	γ^1
Petra	Joan	17	Petra	Garcia	82	100	γ^5
Petra	Joan	17	Johanna	Garcia	18	000	γ^1
Petra	Joan	17	Ricard	Tanaka	18	000	γ^1
J.M.	Casanovas	35	Joana	Casanovas	19	010	γ^3
J.M.	Casanovas	35	Petra	Joan	18	000	γ^1
J.M.	Casanovas	35	J. Manela	Casanovas	35	011	γ^4
J.M.	Casanovas	35	Johanna	Garcia	53	000	γ^1
J.M.	Casanovas	35	Ricard	Garcia	14	000	γ^1
J.M.	Casanovas	35	Petra	Garcia	82	000	γ^1
J.M.	Casanovas	35	Johanna	Garcia	18	000	γ^1
J.M.	Casanovas	35	Ricard	Tanaka	18	000	γ^1
Johanna	Garcia	53	Joana	Casanovas	19	000	γ^1
Johanna	Garcia	53	Petra	Joan	18	000	γ^1
Johanna	Garcia	53	J. Manela	Casanovas	35	000	γ^1
Johanna	Garcia	53	Johanna	Garcia	53	111	γ^8
Johanna	Garcia	53	Ricard	Garcia	14	010	γ^3
Johanna	Garcia	53	Petra	Garcia	82	010	γ^3
Johanna	Garcia	53	Johanna	Garcia	18	110	γ^7
Johanna	Garcia	53	Ricard	Tanaka	18	000	γ^1
Ricardo	Garcia	14	Joana	Casanovas	19	000	γ^1
Ricardo	Garcia	14	Petra	Joan	18	000	γ^1
Ricardo	Garcia	14	J. Manela	Casanovas	35	000	γ^1
Ricardo	Garcia	14	Johanna	Garcia	53	010	γ^3
Ricardo	Garcia	14	Ricard	Garcia	14	011	γ^4

(continued)

Table 5.5 (continued)

Name^A	Surname^A	Age^A	Name^B	Surname^B	Age^B	$\gamma(a,b)$	$\gamma(a,b)$
Ricardo	Garcia	14	Petra	Garcia	82	010	γ^3
Ricardo	Garcia	14	Johanna	Garcia	18	010	γ^3
Ricardo	Garcia	14	Ricard	Tanaka	18	000	γ^1
Petra	Garcia	18	Joana	Casanovas	19	000	γ^1
Petra	Garcia	18	Petra	Joan	18	101	γ^6
Petra	Garcia	18	J. Manela	Casanovas	35	000	γ^1
Petra	Garcia	18	Johanna	Garcia	53	010	γ^3
Petra	Garcia	18	Ricard	Garcia	14	010	γ^3
Petra	Garcia	18	Petra	Garcia	82	110	γ^7
Petra	Garcia	18	Johanna	Garcia	18	011	γ^4
Petra	Garcia	18	Ricard	Tanaka	18	001	γ^2
Johanna	Garcia	18	Joana	Casanovas	19	000	γ^1
Johanna	Garcia	18	Petra	Joan	18	001	γ^2
Johanna	Garcia	18	J. Manela	Casanovas	35	000	γ^1
Johanna	Garcia	18	Johanna	Garcia	53	110	γ^7
Johanna	Garcia	18	Ricard	Garcia	14	010	γ^3
Johanna	Garcia	18	Petra	Garcia	82	010	γ^3
Johanna	Garcia	18	Johanna	Garcia	18	111	γ^8
Johanna	Garcia	18	Ricard	Tanaka	18	001	γ^2
Ricard	Tanaka	14	Joana	Casanovas	19	000	γ^1
Ricard	Tanaka	14	Petra	Joan	17	000	γ^1
Ricard	Tanaka	14	J. Manela	Casanovas	35	000	γ^1
Ricard	Tanaka	14	Johanna	Garcia	53	000	γ^1
Ricard	Tanaka	14	Ricard	Garcia	14	101	γ^6
Ricard	Tanaka	14	Petra	Garcia	82	000	γ^1
Ricard	Tanaka	14	Johanna	Garcia	18	000	γ^1
Ricard	Tanaka	14	Ricard	Tanaka	18	110	γ^7

Remark that, in the above example, $R(\gamma)$ does not really use the values in records a and b but only their coincidences. The rationale of Expression 5.6 is made clear in the rest of this section and the use of log is explained in Sect. 5.5.2. Nevertheless, note that this rule associates large values of R to those pairs whose γ is such that $P(\gamma = \gamma(a',b')|(a',b') \in \mathbf{M})$ is large and $P(\gamma = \gamma(a',b')|(a',b') \in \mathbf{U})$ is small. Therefore, larger values are assigned to $R(\gamma)$ when the probability of finding the coincidence vector γ is larger in $\mathbf{M}$ than in $\mathbf{U}$. Otherwise, small values of R are assigned to coincidence vectors with larger probabilities in $\mathbf{U}$ than in $\mathbf{M}$.

In what follows, we will use m^i and u^i to denote the conditional probabilities of the coincidence vector γ^i:

$$m^i = P(\gamma^i = \gamma(a', b')|(a', b') \in \mathbf{M}) \qquad (5.7)$$

$$u^i = P(\gamma^i = \gamma(a', b')|(a', b') \in \mathbf{U}) \qquad (5.8)$$

Example 5.3 Table 5.6 gives the computation of $R(\gamma^i) = \log(m^i/u^i)$ for all pairs of records in Table 5.5. Probabilities m^i and u^i have been estimated by the proportion of elements in either $\mathbf{M}$ or $\mathbf{U}$ with such coincidence vector γ^i. The table gives coincidence vectors ordered (in decreasing order) according to their R values.

In general, for any decision rule rl, the following two probabilities are of interest:

$$P(\mathbf{LP}|\mathbf{U}) = \mu \qquad (5.9)$$

and

$$P(\mathbf{NP}|\mathbf{M}) = \lambda. \qquad (5.10)$$

Note that they are the probabilities that the rule causes an error. In particular, the first probability corresponds to the classification as a linked pair of a pair that is not a matched pair. This situation corresponds to a *false linkage*. The second probability corresponds to the classification as a non-linked pair of a matched pair. This situation corresponds to a *false unlinkage*. Note that false linkage and false unlinkage can be seen as false positives and false negatives of the classification rule.

Example 5.4 Let (lt, ut) be the lower and upper thresholds used in the decision rule of Definition 5.7. Then, the probabilities $\mu = P(\mathbf{LP}|\mathbf{U})$ and $\lambda = P(\mathbf{NP}|\mathbf{M})$ for this decision rule are equal to:

$$\mu = \sum_{i:\log(m^i/u^i)>ut} u^i$$

$$\lambda = \sum_{i:\log(m^i/u^i)<lt} m^i$$

Assume $lt = 1.5$ and $ut = 2.5$. Using the data from Examples 5.2 and 5.3, we obtain the following values for μ and λ:

$$\mu = 0/56 + 1/56 = 1/56 = 0.0178$$

$$\lambda = 0 + 0 + 0 + 0 + 1/8 = 0.125$$

Table 5.6 Product space $\mathbf{A} \times \mathbf{B}$, Γ vectors, and computation of $R(\gamma^i) = \log(m^i/u^i)$

Name^A	Surname^A	Age^A	Name^B	Surname^B	Age^B	γ^i		M/U	m^i	u^i	m^i/u^i	$\log(m^i/u^i)$
Johanna	Garcia	53	Johanna	Garcia	53	111	γ^8	M	2/8	0/56	∞	∞
Johanna	Garcia	18	Johanna	Garcia	18	111	γ^8	M				
Petra	Joan	17	Petra	Joan	18	110	γ^7	M	3/8	2/56	10.5	2.35
Johanna	Garcia	53	Johanna	Garcia	18	110	γ^7	U				
Petra	Garcia	18	Petra	Garcia	82	110	γ^7	M				
Johanna	Garcia	18	Johanna	Garcia	53	110	γ^7	U				
Ricard	Tanaka	14	Ricard	Tanaka	18	110	γ^7	M				
Joana	Casanoves	19	Joana	Casanovas	19	101	γ^6	M	1/8	2/56	3.5	1.25
Petra	Garcia	18	Petra	Joan	18	101	γ^6	U				
Ricard	Tanaka	14	Ricard	Garcia	14	101	γ^6	U				
Petra	Joan	17	Petra	Garcia	82	100	γ^5	U	0/8	1/56	0	$-\infty$
J.M.	Casanovas	35	J. Manela	Casanovas	35	011	γ^4	M	2/8	1/56	14	2.63
Ricardo	Garcia	14	Ricard	Garcia	14	011	γ^4	M				
Petra	Garcia	18	Johanna	Garcia	18	011	γ^4	U				
Joana	Casanoves	19	J. Manela	Casanovas	35	010	γ^3	U	0/8	11/56	0	$-\infty$
J.M.	Casanovas	35	Joana	Casanovas	19	010	γ^3	U				
Johanna	Garcia	53	Ricard	Garcia	14	010	γ^3	U				
Johanna	Garcia	53	Petra	Garcia	82	010	γ^3	U				
Ricardo	Garcia	14	Johanna	Garcia	53	010	γ^3	U				
Ricardo	Garcia	14	Petra	Garcia	82	010	γ^3	U				
Ricardo	Garcia	14	Johanna	Garcia	18	010	γ^3	U				
Petra	Garcia	18	Petra	Garcia	53	010	γ^3	U				
Petra	Garcia	18	Ricard	Garcia	14	010	γ^3	U				

(continued)

Table 5.6 (continued)

Name^A	Surname^A	Age^A	Name^B	Surname^B	Age^B	γ^i		M/U	m^i	u^i	m^i/u^i	$\log(m^i/u^i)$
Johanna	Garcia	18	Ricard	Garcia	14	010	γ^3	U				
Johanna	Garcia	18	Petra	Garcia	82	010	γ^3	U				
Petra	Garcia	18	Ricard	Tanaka	18	001	γ^2	U	0/8	3/56	0	$-\infty$
Johanna	Garcia	18	Petra	Joan	18	001	γ^2	U				
Johanna	Garcia	18	Ricard	Tanaka	18	001	γ^2	U				
Joana	Casanovas	19	Petra	Joan	18	000	γ^1	U	0/8	36/56	0	$-\infty$
Joana	Casanoves	19	Johanna	Garcia	53	000	γ^1	U				
Joana	Casanoves	19	Ricard	Garcia	14	000	γ^1	U				
Joana	Casanoves	19	Petra	Garcia	82	000	γ^1	U				
Joana	Casanoves	19	Johanna	Garcia	18	000	γ^1	U				
Joana	Casanoves	19	Ricard	Tanaka	18	000	γ^1	U				
Petra	Joan	17	Joana	Casanovas	19	000	γ^1	U				
Petra	Joan	17	J. Manela	Casanovas	35	000	γ^1	U				
Petra	Joan	17	Johanna	Garcia	53	000	γ^1	U				
Petra	Joan	17	Ricard	Garcia	14	000	γ^1	U				
Petra	Joan	17	Johanna	Garcia	18	000	γ^1	U				
Petra	Joan	17	Ricard	Tanaka	18	000	γ^1	U				
J.M.	Casanovas	35	Petra	Joan	18	000	γ^1	U				
J.M.	Casanovas	35	Johanna	Garcia	53	000	γ^1	U				
J.M.	Casanovas	35	Ricard	Garcia	14	000	γ^1	U				
J.M.	Casanovas	35	Petra	Garcia	82	000	γ^1	U				
J.M.	Casanovas	35	Johanna	Garcia	18	000	γ^1	U				

(continued)

Table 5.6 (continued)

Name^A	Surname^A	Age^A	Name^B	Surname^B	Age^B	γ^i	γ^1	M/U	m^i	u^i	m^i/u^i	$\log(m^i/u^i)$
J.M.	Casanovas	35	Ricard	Tanaka	18	000	γ^1	U				
Johanna	Garcia	53	Joana	Casanovas	19	000	γ^1	U				
Johanna	Garcia	53	Petra	Joan	18	000	γ^1	U				
Johanna	Garcia	53	J. Manela	Casanovas	35	000	γ^1	U				
Johanna	Garcia	53	Ricard	Tanaka	18	000	γ^1	U				
Ricardo	Garcia	14	Joana	Casanovas	19	000	γ^1	U				
Ricardo	Garcia	14	Petra	Joan	18	000	γ^1	U				
Ricardo	Garcia	14	J. Manela	Casanovas	35	000	γ^1	U				
Ricardo	Garcia	14	Ricard	Tanaka	18	000	γ^1	U				
Petra	Garcia	18	Joana	Casanovas	19	000	γ^1	U				
Petra	Garcia	18	J. Manela	Casanovas	35	000	γ^1	U				
Johanna	Garcia	18	Joana	Casanovas	19	000	γ^1	U				
Johanna	Garcia	18	J. Manela	Casanovas	35	000	γ^1	U				
Ricard	Tanaka	14	Joana	Casanovas	19	000	γ^1	U				
Ricard	Tanaka	14	Petra	Joan	17	000	γ^1	U				
Ricard	Tanaka	14	J. Manela	Casanovas	35	000	γ^1	U				
Ricard	Tanaka	14	Johanna	Garcia	53	000	γ^1	U				
Ricard	Tanaka	14	Petra	Garcia	82	000	γ^1	U				
Ricard	Tanaka	14	Johanna	Garcia	18	000	γ^1	U				

Table 5.7 Product space $\mathbf{A} \times \mathbf{B}$ and corresponding Γ vectors

Name$^\mathbf{A}$	Surname$^\mathbf{A}$	Age$^\mathbf{A}$	Name$^\mathbf{B}$	Surname$^\mathbf{B}$	Age$^\mathbf{B}$	γ^i		M/U	$m^{\sigma(i)}$	$u^{\sigma(i)}$	$m^{\sigma(i)}/u^{\sigma(i)}$
Johanna ⋮	Garcia	53	Johanna	Garcia	53	111	$\gamma^{\sigma(1)}$	M	2/8	0/56	∞
J.M. ⋮	Casanovas	35	J. Manela	Casanovas	35	011	$\gamma^{\sigma(2)}$	M	2/8	1/56	14
Petra ⋮	Joan	17	Petra	Joan	18	110	$\gamma^{\sigma(3)}$	M	3/8	2/56	10.5
Joana ⋮	Casanoves	19	Joana	Casanovas	19	101	$\gamma^{\sigma(4)}$	M	1/8	2/56	3.5
Petra ⋮	Joan	17	Petra	Garcia	82	100	$\gamma^{\sigma(5)}$	U	0/8	1/56	0
Joana ⋮	Casanoves	19	J. Manela	Casanovas	35	010	$\gamma^{\sigma(6)}$	U	0/8	11/56	0
Petra ⋮	Garcia	18	Ricard	Tanaka	18	001	$\gamma^{\sigma(7)}$	U	0/8	3/56	0
Joana ⋮	Casanoves	19	Petra	Joan	18	000	$\gamma^{\sigma(8)}$	U	0/8	36/56	0

In addition to the two conditional probabilities above, another probability is also relevant in decision rules: the probability of classifying pairs of records into the set **CP**. As this latter set corresponds to pairs that should be further revised, the smaller the probability, the better. Therefore, it is clear that, given the set of all decision rules with the same probabilities $P(\mathbf{LP}|\mathbf{U})$ and $P(\mathbf{NP}|\mathbf{M})$, we are interested in finding the one (or ones) with the smallest probability of classifying a pair as **CP**.

To that end, Fellegi and Sunter [96] considered the following definitions.

Definition 5.8 Let rl be a decision rule in the space Γ and let μ and λ be the two values in the interval $(0, 1)$ for its conditional probabilities $P(\mathbf{LP}|\mathbf{U})$ and $P(\mathbf{NP}|\mathbf{M})$ (Expressions 5.9 and 5.10). Then rl is a *rule with levels μ and λ* and is expressed by $rl(\mu, \lambda, \Gamma)$.

Definition 5.9 Let **rl** be the set of all decision rules over Γ with levels μ and λ. Then $rl(\mu, \lambda, \Gamma)$ is the *optimal decision rule* if it satisfies:

$$P(\mathbf{CP}|rl) \leq P(\mathbf{CP}|rl')$$

for all $rl'(\mu, \lambda, \Gamma)$ in **rl**.

In these definitions, it is assumed that μ and λ lead to a non-empty set of decision rules. It is said that μ and λ are admissible when they satisfy simultaneously Expressions 5.9 and 5.10 and when the set of decision rules is not empty. See [96] for details on the admissibility of μ and λ.

Fellegi and Sunter define an optimal decision rule based on Expression 5.5 and, as will be seen later, the rule is similar to the one we have given in Definition 5.7. This optimal decision rule is defined below.

Definition 5.10 [96] Let μ and λ be an admissible pair of error levels and σ be a permutation of $\{1, \ldots, |\Gamma|\}$ such that $\sigma(j) < \sigma(k)$ if

$$\frac{P(\gamma^{\sigma(j)} = \gamma(a', b')|(a', b') \in \mathbf{M})}{P(\gamma^{\sigma(j)} = \gamma(a', b')|(a', b') \in \mathbf{U})} > \frac{P(\gamma^{\sigma(k)} = \gamma(a', b')|(a', b') \in \mathbf{M})}{P(\gamma^{\sigma(k)} = \gamma(a', b')|(a', b') \in \mathbf{U})} \quad (5.11)$$

and let *limit* and *limit'* be the indexes such that

$$\sum_{i=1}^{limit-1} u^{\sigma(i)} < \mu \leq \sum_{i=1}^{limit} u^{\sigma(i)} \quad (5.12)$$

$$\sum_{i=limit'+1}^{|\Gamma|} m^{\sigma(i)} < \lambda \leq \sum_{i=limit'}^{|\Gamma|} m^{\sigma(i)} \quad (5.13)$$

where u^i and m^i correspond to the conditional probabilities in Expressions 5.7 and 5.8.

Then, the optimal decision rule ODR_p for the pair (a, b) is a probability distribution $(\alpha_1, \alpha_2, \alpha_3)$ on $\{$**LP, CP, NP**$\}$ defined by $ODR_p(a, b) = ODR(\gamma(a, b))$ with ODR defined as follows:

$$ODR(\gamma^{\sigma(i)}) = \begin{cases} (1, 0, 0) & \text{if } 1 \leq i \leq limit - 1 \\ (P_\mu, 1 - P_\mu, 0) & \text{if } i = limit \\ (0, 1, 0) & \text{if } limit < i < limit' \\ (0, 1 - P_\lambda, P_\lambda) & \text{if } i = limit' \\ (0, 0, 1) & \text{if } limit' + 1 \leq i \leq |\Gamma| \end{cases} \qquad (5.14)$$

and where P_μ and P_λ are the solutions of the equations:

$$u^{\sigma(limit)} P_\mu = \mu - \sum_{i=1}^{limit-1} u^{\sigma(i)} \qquad (5.15)$$

$$m^{\sigma(limit')} P_\lambda = \lambda - \sum_{i=limit'+1}^{|\Gamma|} m^{\sigma(i)} \qquad (5.16)$$

This decision rule is optimal. This is established in the next theorem.

Theorem 5.1 *[96] The decision rule in Definition 5.10 is a best decision rule on Γ at the levels μ and λ.*

According to the procedure outlined above, the classification of a pair (a, b) requires: (i) computing the coincidence vector γ; (ii) determining the position of this γ vector in Γ once elements in Γ are ordered according to Expression 5.11 and (iii) computing the probability distribution over **DR** for this $\gamma^{\sigma(i)}$.

5.5.1 Alternative Expressions for Decision Rules

In the particular case that μ and λ satisfy the following equations:

$$\mu = \sum_{i=1,limit} u^{\sigma(i)} \qquad (5.17)$$

$$\lambda = \sum_{i=limit',N_\Gamma} m^{\sigma(i)} \qquad (5.18)$$

the decision rule in Expression 5.14 can be simplified as:

$$SimpODR(\gamma^{\sigma(i)}) = \begin{cases} (1,0,0) \text{ if } 1 \leq i \leq limit \\ (0,1,0) \text{ if } limit < i < limit' \\ (0,0,1) \text{ if } limit' \leq i \leq |\Gamma| \end{cases} \quad (5.19)$$

This rule uses σ, $limit$ and $limit'$ as given in Definition 5.10, and is also optimal under the established conditions for μ and λ in Expressions 5.17 and 5.18.

Nevertheless, when Eqs. 5.17 and 5.18 do not hold, Rule 5.19 is not applicable and the previous definition with probability distributions is needed. To avoid the use of such probability distributions, that make practical applications more complex, we can classify as clerical pairs (assign them to class **CP**) those pairs that lead to a $\gamma^{\sigma(i)}$ with $i = limit$ or $i = limit'$. This is equivalent to using the following decision rule:

$$AltDR(\gamma^{\sigma(i)}) = \begin{cases} (1,0,0) \text{ if } 1 \leq i \leq limit - 1 \\ (0,1,0) \text{ if } limit \leq i \leq limit' \\ (0,0,1) \text{ if } limit' + 1 \leq i \leq |\Gamma| \end{cases} \quad (5.20)$$

Note that in this rule μ is used as an error bound, because the probability $P(\mathbf{LP}|\mathbf{U})$ of the new rule is smaller than the probability of the previous rule and, thus, smaller than μ. This is stated in Proposition 5.2. The same applies for λ (Proposition 5.3).

Proposition 5.2 *Let* $P_{ODR}(\mathbf{LP}|\mathbf{U})$ *and* $P_{AltDR}(\mathbf{LP}|\mathbf{U})$ *be the probabilities of the optimal decision rule ODR in Definition 5.10 and of the decision rule in Eq. 5.20. Then,*

$$P_{AltDR}(\mathbf{LP}|\mathbf{U}) < P_{ODR}(\mathbf{LP}|\mathbf{U}) = \mu$$

Proposition 5.3 *Let* $P_{ODR}(\mathbf{NP}|\mathbf{M})$ *and* $P_{AltDR}(\mathbf{NP}|\mathbf{M})$ *be the probabilities of the optimal decision rule ODR in Definition 5.10 and of the decision rule in Eq. 5.20. Then,*

$$P_{AltDR}(\mathbf{NP}|\mathbf{M}) < P_{ODR}(\mathbf{NP}|\mathbf{M}) = \lambda$$

Nevertheless, the rule in Expression 5.20 also classifies as clerical pairs those pairs (a, b) with $\gamma(a, b) = \gamma^{\sigma(limit)}$ or $\gamma(a, b) = \gamma^{\sigma(limit')}$ when Expressions 5.17 and 5.18 hold. Since these pairs can be classified as **LP** and **NP** without violating the bounds

$$P(\mathbf{LP}|\mathbf{U}) \leq \mu$$

and

$$P(\mathbf{NP}|\mathbf{M}) \leq \lambda$$

the previous rule can be rewritten as follows:

$$rule(\gamma^{\sigma(i)}) = \begin{cases} (1, 0, 0) \text{ if } 1 \leq i \leq limit \\ (0, 1, 0) \text{ if } limit < i < limit' \\ (0, 0, 1) \text{ if } limit' \leq i \leq N_\Gamma \end{cases} \quad (5.21)$$

This requires that the indices *limit* and *limit'* be determined according to the following inequalities:

$$\sum_{i=1}^{limit} u^{\sigma(i)} \leq \mu < \sum_{i=1}^{limit+1} u^{\sigma(i)} \quad (5.22)$$

$$\sum_{i=limit'}^{|\Gamma|} m^{\sigma(i)} \leq \lambda < \sum_{i=limit'-1}^{|\Gamma|} m^{\sigma(i)} \quad (5.23)$$

It is important to underline that these latter rules are non-optimal rules because the probability of classifying a pair as a clerical pair is larger than the one in Definition 5.10. However, from a practical point of view, the last rule is convenient and easy to use; for example, it was the rule used in [85].

For the application of the decision rules defined so far, we need to know the position of the coincidence vector $\gamma^{\sigma(i)}$ in the ordering obtained from Γ and also the indexes *limit* and *limit'*. We give below an alternative definition that does not require these elements. This definition is equivalent to the one given above when appropriate thresholds are selected. This rule corresponds to the one presented in Definition 5.7.

Definition 5.11 Let (a, b) be a pair of records in $\mathbf{A} \times \mathbf{B}$, let (lt, ut) be two thresholds (lower and upper) in $\mathbb{R}$ such that $lt < ut$, then the *Decision Rule* is defined as follows:

1. If $R_p(a, b) \geq ut$ then (a, b) is a Linked Pair (**LP**)
2. If $R_p(a, b) \leq lt$ then (a, b) is a Non linked Pair (**NP**)
3. If $lt < R_p(a, b) < ut$ then (a, b) is a Clerical Pair (**CP**)

where the index $R_p(a, b)$ is defined in terms of the vector of coincidences $\gamma(a, b)$ using Eq. 5.5 as follows

$$R_p(a, b) = R(\gamma(a, b)) = \log\left(\frac{P(\gamma(a, b) = \gamma(a', b')|(a', b') \in \mathbf{M})}{P(\gamma(a, b) = \gamma(a', b')|(a', b') \in \mathbf{U})}\right). \quad (5.24)$$

Proposition 5.4 *The decision rule in Definition 5.11 is equivalent to the decision rule in Expression 5.21 when lt and ut are defined as follows:*

$$ut = \log(m^{\sigma\,(limit)}/u^{\sigma\,(limit)})$$

$$lt = \log(m^{\sigma\,(limit')}/u^{\sigma\,(limit')})$$

where, as usual, $m^i = P(\gamma^i = \gamma(a', b')|(a', b') \in \mathbf{M})$, $u^i = P(\gamma^i = \gamma(a', b')|$ $(a', b') \in \mathbf{U})$.

Example 5.5 Let us consider the probabilistic record linkage of records defined in terms of the three variables (*Name, Surname, Age*) as in Example 5.2. Let us consider the conditional probabilities m^i and u^i inferred from files **A** and **B** in Example 5.2 and computed in Example 5.3 (displayed in Table 5.6).

Now, let us compute the decision rule for $\mu = 0.05$ and $\lambda = 0.2$ and show its application to classify the pair ((*J. Gomez, 19*), (*P. Gomez, 19*)).

First, to define the rule, we need to determine *limit* and *limit'* to apply Proposition 5.4. These values are set by Expressions 5.22 and 5.23 and the conditional probabilities in Table 5.6. Taking all this into account, we get

$$\sum_{i=1}^{2} u^{\sigma\,(i)} = 0 + 0.017 \le 0.05 < 0 + 0.017 + 0.035 = \sum_{i=1}^{2+1} u^{\sigma\,(i)}$$

and

$$\sum_{i=5}^{|\Gamma|} m^{\sigma\,(i)} = 0 + 0 + 0 + 0 + 0.125 \le 0.2 < 0 + 0 + 0 + 0 + 0.125 + 0.375 = \sum_{i=5-1}^{|\Gamma|} m^{\sigma\,(i)}.$$

Therefore, *limit* $= 2$ and *limit'* $= 5$. Thus, $ut = \log(m^{\sigma\,(limit)}/u^{\sigma\,(limit)}) = \log(14) = 2.63$ and $lt = \log(m^{\sigma\,(limit')}/u^{\sigma\,(limit')}) = \log(3.5) = 1.25$.

According to this, the rule becomes:

1. If $R_p(a, b) \ge 2.63$ then (a, b) is a Linked Pair
2. If $R_p(a, b) \le 1.25$ then (a, b) is a Non linked Pair
3. If $lt < R_p(a, b) < ut$ then (a, b) is a Clerical Pair

Now, we can consider any pair of records and classify them using this rule. If we take the pair (*J.Gomez*19), (*P.Gomez*19) we first compute the coincidence vector γ. We get $\gamma = (011)$. Then, we need to compute for this vector $R(011)$. Using the values m^i and u^i in Table 5.6 we get $R(011) = \log(14) = 2.63$ (see Table 5.7). Therefore, the rule classifies the pair as a linked pair.

In this section we have seen how to define the decision rule and how to apply it to a pair of records. However, this process requires several conditional probabilities to be determined. One possibility is to start with a pair of files for which the matched pairs are known (as in the examples in this section) and then estimate the probabilities by proportions of records. In the next sections we consider in more detail the computation of $R(a, b)$ and the estimation of the probabilities involved in $R(a, b)$.

5.5.2 Computation of $R_p(a, b)$

Some general aspects about the computation of $R_p(a, b)$ for a given pair (a, b) in
$\mathbf{A} \times \mathbf{B}$ are described in this section. Specifically, the estimation of the probabilities
involved in this computation is detailed in Sect. 5.5.3. In fact, according to the rule,
the computation of $R_p(a, b)$ is solely based on the computation of R for $\gamma(a, b)$.

Due to the fact that the cardinality of Γ is typically quite large (recall that $|\Gamma| = 2^n$
where n is the number of variables), it is not appropriate to directly estimate the probabilities of all γ. To avoid this computation, it is usual to assume that the components
of the vector $\gamma = (\gamma_1, \ldots, \gamma_n)$ are statistically independent. Under this assumption,
the probabilities $P(\gamma = \gamma(a, b)|(a, b) \in \mathbf{M})$ and $P(\gamma = \gamma(a, b)|(a, b) \in \mathbf{U})$ can be
expressed in the following way:

$$P(\gamma = \gamma(a, b)|(a, b) \in \mathbf{M}) = \prod_{i=1,n} P(\gamma_i = \gamma_i(a, b)|(a, b) \in \mathbf{M}) \qquad (5.25)$$

$$P(\gamma = \gamma(a, b)|(a, b) \in \mathbf{U}) = \prod_{i=1,n} P(\gamma_i = \gamma_i(a, b)|(a, b) \in \mathbf{U}) \qquad (5.26)$$

To simplify the notation, we shall use the following equivalences:

- $m(\gamma) = P(\gamma = \gamma(a, b)|(a, b) \in \mathbf{M})$
- $m_i(\gamma_i) = P(\gamma_i = \gamma_i(a, b)|(a, b) \in \mathbf{M})$
- $u(\gamma) = P(\gamma = \gamma(a, b)|(a, b) \in \mathbf{U})$
- $u_i(\gamma_i) = P(\gamma_i = \gamma_i(a, b)|(a, b) \in \mathbf{U})$

Using these equivalences, Eqs. 5.25 and 5.26 are rewritten as:

$$m(\gamma) = \prod_{i=1,n} m_i(\gamma_i)$$

$$u(\gamma) = \prod_{i=1,n} u_i(\gamma_i)$$

Therefore, under the same conditions of independence $R_p(a, b)$ can be rewritten as:

$$R_p(a, b) = R(\gamma(a, b)) = \log\left(\frac{P(\gamma(a, b) = \gamma(a', b')|(a', b') \in \mathbf{M})}{P(\gamma(a, b) = \gamma(a', b')|(a', b') \in \mathbf{U})}\right) \quad (5.27)$$

$$= \log\left(\frac{m(\gamma(a, b))}{u(\gamma(a, b))}\right) \qquad (5.28)$$

$$= \log\left(\frac{\prod_{i=1,n} m_i(\gamma_i(a, b))}{\prod_{i=1,n} u_i(\gamma_i(a, b))}\right) \qquad (5.29)$$

$$= \sum_{i=1,n} \log(m_i(\gamma_i(a, b))/u_i(\gamma_i(a, b))) \qquad (5.30)$$

Note that the use of the logarithm in $R(\gamma)$ simplifies its expression. Now, using that the following expressions about conditional probabilities hold for all $i \in \{1, \ldots, n\}$

$$P(\gamma_i = 1 | (a, b) \in \mathbf{M}) + P(\gamma_i = 0 | (a, b) \in \mathbf{M}) = m_i(1) + m_i(0) = 1$$

$$P(\gamma_i = 1 | (a, b) \in \mathbf{U}) + P(\gamma_i = 0 | (a, b) \in \mathbf{U}) = u_i(1) + u_i(0) = 1$$

we define

- $m_i = m_i(1)$
- $u_i = u_i(1)$

and express $m_i(0)$ and $u_i(0)$ as:

$$m_i(0) = 1 - m_i$$

$$u_i(0) = 1 - u_i$$

These definitions permits us to express the conditional probabilities in an alternative and more compact way (note that here γ_i and $1 - \gamma_i$ is either 1 or 0):

$$P(\gamma = \gamma(a', b') | (a', b') \in \mathbf{M}) = \prod m_i^{\gamma_i} (1 - m_i)^{1-\gamma_i}$$

$$P(\gamma = \gamma(a', b') | (a', b') \in \mathbf{U}) = \prod u_i^{\gamma_i} (1 - u_i)^{1-\gamma_i}.$$

By further defining $w_i(\gamma_i)$ as $\log(m_i(\gamma_i)/u_i(\gamma_i))$, we have that $R(a, b)$ in Expression 5.30 can be rewritten as:

$$R_p(a, b) = R(\gamma(a, b)) = \sum_{i=1}^{n} w_i(\gamma_i(a, b)). \tag{5.31}$$

Thanks to the above definitions, we only need m^i and u^i to compute $w_i(\gamma_i(a, b))$. To do so, two cases are considered:

- Case $\gamma_i(a, b) = 1$: define $w_i(1) = \log(m_i/u_i)$.
- Case $\gamma_i(a, b) = 0$: define $w_i(0) = \log((1 - m_i)/(1 - u_i))$.

Note that these expressions are correct because $w_i(1)$ is equal to $\log(m_i(1)/u_i(1))$ and $m_i = m_i(1)$ and $u_i = u_i(0)$. Similarly, $w_i(0)$ is equal to $\log(m_i(0)/u_i(0))$ and, thus, considering the equalities $m_i(0) = 1 - m_i$ and $u_i(0) = 1 - u_i$ we get the expression above.

The terms $w_i(\gamma_i(a, b))$ are known as the *weights* of $\gamma_i(a, b)$. As the usual case is to have $m_i > u_i$, then, the variables with coincident values (i.e. with $\gamma_i = 1$) contribute positively to the value $R(\gamma)$. Instead, variables with non-coincident values (i.e., with $\gamma_i = 0$) contribute negatively to the value $R(\gamma)$.

In fact, expressions for $w_i(1)$ and $w_i(0)$ given above clearly show that, when $m_i > u_i$, the weights for $\gamma_i(a, b) = 1$ are positive (and thus contribute positively to $R_p(a, b)$) and the weights for $\gamma_i(a, b) = 0$ are negative (and thus contribute negatively to $R_p(a, b)$). This is stated in the next proposition.

Proposition 5.5 *Let $m_i > u_i$, then $w_i(1) > 0$ and $w_i(0) < 0$.*

Proof $m_i > u_i$ implies $m_i/u_i > 1$, therefore $w_i(1) = \log \frac{m_i}{u_i} > \log 1 = 0$. Also, it implies $1 - u_i > 1 - m_i > 0$, therefore, $1 > \frac{1-m_i}{1-u_i}$ and, thus, $0 = \log 1 > \log \frac{1-m_i}{1-u_i}$

Proposition 5.6 *Let $m_i < u_i$, then $w_i(1) < 0$ and $w_i(0) > 0$.*

Let us now turn into the estimation of probabilities m_i and u_i for all $i \in \{1, \ldots, n\}$.

5.5.3 Estimation of the Probabilities

The estimation of the probabilities involved in the computation of R_p is usually based on the EM (Expectation-Maximization) algorithm [97]. We described the EM algorithm in Sect. 2.3.3. In this section, we describe the application of the EM algorithm to the record linkage process.

5.5.3.1 EM Algorithm for Record Linkage

The application of EM to record linkage relies on the consideration of pairs of vectors $< \gamma(r), c(r) >_{r \in \mathbf{A} \times \mathbf{B}}$ as the complete data. Here, as usual, γ is the coincidence vector for $r \in \mathbf{A} \times \mathbf{B}$ and c (c for class) is a two dimensional vector $c = (c_m c_u)$ in $\{(10), (01)\}$ to indicate whether r belongs to $\mathbf{M}$ or $\mathbf{U}$. Then, for all pairs of records r in $\mathbf{A} \times \mathbf{B}$ we consider $(\gamma(r), c(r))$ where $c(r) = (10)$ if and only if $r \in \mathbf{M}$ and $c(r) = (01)$ if and only if $r \in \mathbf{U}$.

Incomplete data correspond to the case that some vectors c are unknown for some records r. Then, the expectation step assigns to the missing indicators fractions that sum to unity that are expectations given the current estimate of the parameters.

Here, the parameters θ consist of probabilities

$$m = (m_1, \ldots, m_n) \text{ and } u = (u_1, \ldots, u_n)$$

with $m_i = P(1 = \gamma_i(a, b)|(a, b) \in \mathbf{M})$ and $u_i = P(1 = \gamma_i(a, b)|(a, b) \in \mathbf{U})$ (as defined in Sect. 5.5.2). Additionally, the parameters θ also contains p (the proportion of matched pairs $p = |M|/|M \cup U|$). Therefore, $\theta = (m, u, p)$.

Then, the log-likelihood for the complete data corresponds to [85,97]:

$$\ln f(\mathbf{x}|\theta) = \sum_{j=1}^{N} c(r^j)(\ln P\{\gamma(r^j)|\mathbf{M}\}, \ln P\{\gamma(r^j)|\mathbf{U}\})^T + \sum_{j=1}^{N} c(r^j)(\ln p, \ln(1-p))^T$$

This expression allows us to estimate the probabilities of assigning records in $\mathbf{A} \times \mathbf{B}$ either to the class $\mathbf{M}$ or $\mathbf{U}$ once an estimation for the parameters $\theta = (m, u, p)$ is given. In fact, the assignment does only depend on the corresponding coincidence vector. Therefore, the estimation is computed for the coincidence vectors $\gamma^j \in \Gamma$. This is the expectation step (see [85]), which yields the following assignment probabilities:

$$\hat{c}_m(\gamma^j) = \frac{\hat{p} \prod_{i=1}^{n} \hat{m}_i^{\gamma_i^j} (1-\hat{m}_i)^{1-\gamma_i^j}}{\hat{p} \prod_{i=1}^{n} \hat{m}_i^{\gamma_i^j} (1-\hat{m}_i)^{1-\gamma_i^j} + (1-\hat{p}) \prod_{i=1}^{n} \hat{u}_i^{\gamma_i^j} (1-\hat{u}_i)^{(1-\gamma_i^j)}} \quad (5.32)$$

$$\hat{c}_u(\gamma^j) = \frac{(1-\hat{p}) \prod_{i=1}^{n} \hat{u}_i^{\gamma_i^j} (1-\hat{u}_i)^{1-\gamma_i^j}}{\hat{p} \prod_{i=1}^{n} \hat{m}_i^{\gamma_i^j} (1-\hat{m}_i)^{1-\gamma_i^j} + (1-\hat{p}) \prod_{i=1}^{n} \hat{u}_i^{\gamma_i^j} (1-\hat{u}_i)^{(1-\gamma_i^j)}} \quad (5.33)$$

Then, in the maximization step, we need to calculate a new estimation of the parameters in θ. Therefore, we compute $\hat{m}_i$, $\hat{u}_i$ for all variables $i \in \{1, ..., n\}$ and recompute $\hat{p}$. This is done using the following equations (see [85]):

$$\hat{m}_i = \frac{\sum_{j=1}^{N} [\hat{c}_m(\gamma(r^j))\gamma_i(r^j)]}{\sum_{j=1}^{N} [\hat{c}_m(\gamma(r^j))]} \quad (5.34)$$

$$\hat{u}_i = \frac{\sum_{j=1}^{N} [\hat{c}_u(\gamma(r^j))\gamma_i(r^j)]}{\sum_{j=1}^{N} [\hat{c}_u(\gamma(r^j))]} \quad (5.35)$$

$$\hat{p} = \frac{\sum_{j=1}^{N} [\hat{c}_m(\gamma(r^j))]}{N} \quad (5.36)$$

Although the latter equations are written to consider all pairs of records in $\mathbf{A} \times \mathbf{B}$, it is advisable to accumulate the frequencies of each coincidence vector γ and use alternative expressions. If $fq(\gamma^j)$ is the frequency of the γ^j coincidence vector, then equations above for $\hat{m}_i$, $\hat{u}_i$ and $\hat{p}$ can be rewritten as follows.

$$\hat{m}_i = \frac{\sum_{j=1}^{2^n} [\hat{c}_m(\gamma^j)\gamma_i^j fq(\gamma^j)]}{\sum_{j=1}^{2^n} [\hat{c}_m(\gamma^j) fq(\gamma^j)]} \quad (5.37)$$

$$\hat{u}_i = \frac{\sum_{j=1}^{2^n}[\hat{c}_u(\gamma^j)\gamma_i^j fq(\gamma^j)]}{\sum_{j=1}^{2^n}[\hat{c}_u(\gamma^j)fq(\gamma^j)]} \tag{5.38}$$

$$\hat{p} = \frac{\sum_{j=1}^{2^n}[\hat{c}_m(\gamma^j)fq(\gamma^j)]}{\sum_{i=1}^{2^n} fq(\gamma^j)} \tag{5.39}$$

5.5.3.2　Initialization Step

In [85], it is stated that the algorithm is not very sensitive to initial values for m and u, although values $m_i > u_i$ are advisable. $m_i = 0.9$ was reported in [85] and $m_i = 0.9$ and $u_i = 0.1$ were used in [18,98].

5.5.4　Extensions for Computing Probabilities

In this section we discuss some variations that can be found in the literature to estimate the probabilities.

5.5.4.1　Considering Relative Frequencies

It is well-known that for a given variable, not all values are equally probable. For example, in any country there are common surnames and not so common ones. E.g., e.g. *Garcia* is common in Barcelona and in Table 5.4 but *Tanaka* is not. Then, it is natural to consider models in which the conditional probabilities are influenced by the values. Newcombe et al. [99] already considered this situation and latter on mathematical models were formulated by e.g. Fellegi and Sunter [96] and Winkler [72].

Their approach is based on the fact that coincidence on a particular variable is due to the coincidence on a particular value of that variable. Therefore, the probability of coincidence on the variable can be expressed on the probabilities of coincidence on the values in that variables' range. Therefore, for a variable V_i with $D(V_i) = \{l_1^i, ..., l_{n_{V_i}}^i\}$, the following holds:

$$P(\text{agree on variable } V_i|\mathbf{M}) = \sum_k P(\text{agree on term } k \text{ of variable } V_i|\mathbf{M})$$

or, equivalently,

$$m_i(\gamma_i) = P(\gamma_i(a, b)|\mathbf{M}) = \sum P(V_i(a) = V_i(b) = l_k^i|\mathbf{M}) \tag{5.40}$$

According to Fellegi and Sunter [96] (Section 3.3.1) and [72] we denote the true frequencies of a string in the files A and B by:

$$f_i = |\{a \in A | V_i(a) = l_k^i\}|$$
$$g_i = |\{b \in B | V_i(b) = l_k^i\}|$$

The sum of frequencies (the number of records) in A and B is:

$$\sum f_i = N_A$$
$$\sum g_i = N_B$$

Then, the frequencies for the intersection of the two files is

$$h_1, h_2, \ldots, h_m$$

where $\sum_i h_i = N_{AB}$. It follows that $h_i \leq \min(f_i, g_i)$ as for each record in A we can have at most a correct link between A and B. Reference [72] states that in some implementations

$$P(\text{agree on term } k \text{ of variable } V_i | \mathbf{M}) = h_k / N_{AB}$$

and

$$P(\text{agree on term } k \text{ of variable } V_i | \mathbf{U}) = \frac{f_k g_k - h_k}{N_A N_B - N_{AB}}.$$

The estimation of $m_i(\gamma_i) = P(\text{agree on variable } V_i | \mathbf{M})$ is done using the EM algorithm. For estimating f_i, g_i, observed values must be used and for h_i they use the minimum (i.e., $h_i = \min(f_i, g_i)$). Then, Eq. 5.40 is used for the scaling. That is, $P(\text{agree on term } k \text{ of variable } V_i | \mathbf{M}) = \kappa h_k / \sum_i h_i$ where κ is the estimation obtained from the EM algorithm for $m_i(\gamma_i)$.

5.5.4.2 Using Blocking for Estimating Probabilities

The EM algorithm has been applied in [85] storing frequency counts withing blocks to reduce computation. Then, pairs of records (a, b) are only considered within the blocks (same values for the blocking variables). Coincidences are examined (i.e., the vector γ is computed) and counts are updated (i.e., 1 is added to the frequency count for that particular γ). Counters are accumulated for all blocks. In this way, counts represent the number of observed configurations over all blocks. It is important to stress that such simplification leads to different counts. In fact, blocking reduces the number of unmatched pairs and, therefore, the probabilities u_i will be underestimated. To avoid underestimation, the probability

$$u_i = P(1 = \gamma_i(a, b) | (a, b) \in U)$$

is estimated by:

$$\hat{u}_i = P(1 = \gamma_i(a, b)).$$

The above probability is the probability of $V_i(a) = V_i(b)$ and can be computed directly from the files **A** and **B** counting, for example, the number of pairs with the same value for the variable V_i:

$$\hat{u}_i = \frac{|\{V_i(a) = V_i(b)\}|}{N}.$$

5.5.4.3   Adapting Weights w_i

If partial coincidence is considered when comparing variable values, the weights attached to variables (w_i following the notation in Sect. 5.5.2) must be updated according to the coincidence. Usually, the update is proportional to the similarity. For example, multiplying the weight by the similarity:

$$w_i'(\gamma_i(a, b)) = w_i(\gamma_i(a, b)) \cdot similarity(V_i(a), V_i(b)).$$

Moreover, to improve the performance of the method, updating the similarity function by applying a particular transformation f is sometimes required. Therefore, an expression similar to the following one is used:

$$w_i'(\gamma_i(a, b)) = w_i(\gamma_i(a, b)) \cdot f(similarity(V_i(a), V_i(b)))$$

A particular example of the transformation function is the one used in [89]:

$$f(x) = \begin{cases} x^{0.2435} & \text{if } x > 0.8 \\ 0.0 & \text{if } x \le 0.8 \end{cases}$$

When a file with known coincidences is available, it is possible to learn these functions from the examples in that file.

5.5.4.4   The Independence Assumption

Most literature on probabilistic record linkage assumes the independence of the variables. This is the approach considered in our description. Recall Sect. 5.5.2 and Eqs. 5.25 and 5.26. Nevertheless, it is usual that some variables are not independent. Consider e.g., postal code and city of residence.

Winkler [100] discusses the problem of record linkage in the case of non-independent variables introducing interaction models. In each interaction model, a set of interaction patterns are considered. The paper compares two interaction models, the one in [101] and the one in [102]. Tromp et al. [103] also discuss the case of non independent variables. They revise the expressions we have given in our Sect. 5.5.2 to cope with dependencies between pairs of variables. Finally, [104] also introduces a model that includes terms to model dependences between pairs of fields.

Algorithm 17: Distance-based record linkage.

Data: A: file; B: file
Result: LP: linked pairs; NP: non-linked pairs

1 **for** $a \in A$ **do**
2   b' = arg $\min_{b \in B} d(a, b)$;
3     $LP = LP \cup (a, b')$;
4   **for** $b \in B$ **such that** $b \neq b'$ **do**
5        $NP := NP \cup (a, b)$;

6 **return** (LP, NP) ;

5.5.5 Final Notes

This section was based on [95] that describes probabilistic record linkage as used for our own implementation in Java. It was based on [85,96].

For further reading on probabilistic record linkage, see e.g. the book by Herzog, Scheuren, and Winkler [105] and the reviews [106–108].

Protocols to ensure private probabilistic record linkage have also been developed. See e.g. [109] for a survey and references.

5.6 Distance-Based Record Linkage

This approach, which was first described in [110] in a very restricted formulation for disclosure risk assessment, consists of computing distances between records in the two data files being considered. Then, the pair of records at minimum distance are considered linked pairs.

Let $d(a, b)$ be a distance function between records in file **A** and file **B**. Then, Algorithm 17 describes distance-based record linkage. Naturally, the effectiveness of this record linkage algorithm heavily relies on the effectiveness of the distance function.

We have seen in Sect. 5.4.7 the basics for comparing records. We have seen some distances for different types of variables and Definition 5.2 to compute distance between pairs of records.

Definition 5.2 assumes that all variables are equally important and that they are all independent. We will consider now weighted distances that can be used when these assumptions fail. Weighted distances are necessary if we want to distinguish the variables that have the best discriminatory power for record linkage.

5.6.1   Weighted Distances

Let us reconsider Definition 5.2. For the sake of simplicity we use the square of the distance.

Note that the square of the distance is not really a distance (as it does not satisfy the triangular inequality) but from a practical point of view there is no difference. Note that replacing in Algorithm 17 d by its square d^2 and/or multiplying d by a constant will result in exactly the same sets (LP, NP). Because of that we consider the following expression.

Definition 5.12 Assuming equal weight for all variables $\mathbf{V} = \{V_1, V_2, \cdots V_n\}$, the square distance between records a and b is defined by:

$$d^2(a, b) = \frac{1}{n} \sum_{i=1}^{n} d_{V_i}(V_i^A(a), V_i^B(b)).$$

Some weighted generalizations of this expression have been considered in the literature. Generalizations replace the implicit arithmetic mean in the previous expression by other aggregation functions [60, 111, 112] as the weighted mean, the OWA and the Choquet integral. Note that if $AM(a_1, \ldots, a_n) = \sum a_i/n$ represents the arithmetic mean, then

$$d^2(a, b) = AM(d_{V_1}(V_1^A(a), V_1^B(b)), \ldots, d_{V_n}(V_n^A(a), V_n^B(b))). \tag{5.41}$$

The weighted mean and the OWA operators use a weighting vector to represent the importance in the variables/data. A vector $w = (w_1, \ldots, w_n)$ is a weighting vector of dimension n if $w_i \geq 0$ and $\sum w_i = 1$. The Choquet integral uses a non-additive measure (also known as a fuzzy measure) to represent the importance of the variables/data. A non-additive measure μ is a set function on the set of variables $\mathbf{V}$ such that $\mu(\emptyset) = 0$, $\mu(\mathbf{V}) = 1$ and for all $A, B \subseteq \mathbf{V}$ such that $A \subseteq B$ we have that $\mu(A) \leq \mu(B)$.

Definition 5.13 Given $a_1, \ldots, a_n \in \mathbb{R}$ the weighted mean (WM), the ordered weighted averaging (OWA) operator, and the Choquet integral (CI) are defined as follows.

- Let **p** be a weighting vector of dimension n; then, a mapping WM: $\mathbb{R}^n \to \mathbb{R}$ is a *weighted mean* of dimension n if $WM_p(a_1, \ldots, a_n) = \sum_{i=1}^{n} p_i a_i$.
- Let **w** be a weighting vector of dimension n; then, a mapping OWA: $\mathbb{R}^n \to \mathbb{R}$ is an *Ordered Weighting Averaging (OWA) operator* of dimension n if

$$OWA_\mathbf{w}(a_1, \ldots, a_n) = \sum_{i=1}^{n} w_i a_{\sigma(i)},$$

where $\{\sigma(1), \ldots, \sigma(n)\}$ is a permutation of $\{1, \ldots, n\}$ such that $a_{\sigma(i-1)} \geq a_{\sigma(i)}$ for all $i = \{2, \ldots, n\}$ (i.e., $a_{\sigma(i)}$ is the ith largest element in the collection $a_1, \ldots, a_n$). Note that the OWA operator is a liner combination of order statistics.

- Let μ be a fuzzy measure on $\mathbf{V}$; then, the *Choquet integral* of a function $f : \mathbf{V} \to \mathbb{R}^+$ with respect to the fuzzy measure μ is defined by

$$(C) \int f d\mu = \sum_{i=1}^{n} [f(x_{s(i)}) - f(x_{s(i-1)})]\mu(A_{s(i)}), \qquad (5.42)$$

where $f(x_{s(i)})$ indicates that the indices have been permuted so that $0 \leq f(x_{s(1)}) \leq \cdots \leq f(x_{s(n)}) \leq 1$, and where $f(x_{s(0)}) = 0$ and $A_{s(i)} = \{x_{s(i)}, \ldots, x_{s(n)}\}$. When no confusion exists over the domain $\mathbf{V}$, we use the notation

$$CI_\mu(a_1, \ldots, a_n) = (C) \int f d\mu,$$

where, $f(x_i) = a_i$, as before.

These definitions permits us to introduce the following distances.

Definition 5.14 Given records a and b on variables $\mathbf{V} = \{V_1, V_2, \cdots V_n\}$, and weighting vectors p, w and a non-additive measure μ on $\mathbf{V}$ we define the following (square) distances:

- $d^2 WM_p(a, b) = WM_p(d_1(a, b), \ldots, d_n(a, b))$,
- $d^2 OWA_w(a, b) = OWA_w(d_1(a, b), \ldots, d_n(a, b))$,
- $d^2 CI_\mu(a, b) = CI_\mu(d_1(a, b), \ldots, d_n(a, b))$,

where $d_i(a, b)$ is defined as $d_{V_i}(V_i^A(a), V_i^B(b))$.

Note that although we are using here the term distance, these expressions are not always, properly speaking, a distance because the triangular inequality is not satisfied.

Due to the properties of aggregation functions (see [60]), $d^2 WM$ and $d^2 OWA$ are generalizations of $d^2 AM$. That is, they are more flexible than $d^2 AM$ and all results achieved by $d^2 AM$ can be achieved by the others with appropriate parameters p and w. Similarly, $d^2 CI$ is more general than $d^2 AM$, $d^2 WM$, and $d^2 OWA$. Thus, using the distance $d^2 CI$ we can obtain at least as much correct reidentifications as with the other distances but may be more.

The difficulty of using these expressions is that we have to tune the parameters. That is, we have to know how to assess the importance of the different variables, or the importance of the different sets of variables in the reidentification process. In any case, when $w = p = (1/n, \ldots, 1/n)$ and $\mu(A) = |A|/n$ we have that all distances are equivalent. This corresponds to state that all variables have the same importance, as implicitly stated in the Euclidean distance.

We have stated that these distances permit us to represent different importance of the variables. In addition, the distance based on the Choquet integral permits to represent situations in which there is no independence between the variables. The measure μ permits to express the interactions. See e.g. [60] for details on the interpretation of non-additive measures. Nevertheless, this is not the only possible approach for considering non independent variables. Another one is the Mahalanobis distance and, in general, any symmetric bilinear form.

In symmetric bilinear forms, the importance or weights will be represented by a symmetric matrix. That is, given a set of n variables, we consider a $n \times n$ symmetric matrix. Recall that a symmetric matrix is one where $a_{ij} = a_{ji}$ for all i, j.

Definition 5.15 Given records a and b on variables $\mathbf{V} = \{V_1, V_2, \cdots V_n\}$, and a $n \times n$ symmetric matrix Q, we define the square of a symmetric bilinear form as

- $d^2 SB_Q(a, b) = SB_Q(d_1(a, b), \ldots, d_n(a, b))$

where

$$SB_Q(c_1, ..., c_n) = (c_1, ..., c_n)' Q^{-1} (c_1, ..., c_n),$$

with c' being the transpose of c, and where $d_i(a, b)$ is defined as $d_{V_i}(V_i^A(a), V_i^B(b))$.

A symmetric matrix M is said to be positive definite if $x'Mx > 0$ for all non zero vectors x, and positive semi-definite if $x'Mx \geq 0$ for all vectors x.

When the matrix is positive definite we have that the root square of the bilinear form is a distance satisfying the triangular inequality, and also the identity of indiscernibles. For positive semi-definite matrices the triangular inequality is satisfied but the identity of indiscernibles is not.

When Q is the covariance matrix (i.e., $Q = \Sigma$ when Σ is the covariance matrix of the data), which is semi-definite positive, we have that Definition 5.15 leads to the Mahalanobis distance. That is, $\sqrt{d^2 SB_\Sigma(a, b)}$ is the Mahalanobis distance between a and b for the covariance matrix Σ.

Recall that the covariance matrix Σ is computed by

$$[Var(V^X) + Var(V^Y) - 2Cov(V^X, V^Y)],$$

where $Var(V^X)$ is the variance of variables V^X, $Var(V^Y)$ is the variance of variables V^Y and $Cov(V^X, V^Y)$ is the covariance between variables V^X and V^Y.

The bilinear form generalizes the Euclidean distance when Q^{-1} is diagonal and such that $a_{ii} = 1/n$ and generalizes the weighted mean when Q^{-1} is diagonal and such that $a_{ii} = w_i$. However, in all other cases $d^2 SB$ and $d^2 CI$ lead to different results. Both distances permit to represent situations in which there is no independence between the variables but the type of relationship between the variables is different.

All these expressions have been used to evaluate disclosure risk. The Mahalanobis distance in [27]. The distance based on the weighted means in [113,114], the one based on the OWA in [114], and the one based on the Choquet integral in [115].

Kernel distances have also been considered (see e.g. [27]). Their definition is for numerical data. The main idea is that instead of computing the distance between records a and b in the original n dimensional space, records are compared in a higher dimensional space H. The definition follows.

Definition 5.16 Let $\Phi(x)$ be the mapping of x into the higher space. Then, the distance between records a and b in H is defined as follows:

$$
\begin{aligned}
d_K^2(a, b) = \|\Phi(a) - \Phi(b)\|^2 &= (\Phi(a) - \Phi(b))^2 \\
&= \Phi(a) \cdot \Phi(a) - 2\Phi(a) \cdot \Phi(b) + \Phi(b) \cdot \Phi(b) \\
&= K(a, a) - 2K(a, b) + K(b, b)
\end{aligned}
$$

where K is a kernel function (i.e., $K(a, b) = \Phi(a) \cdot \Phi(b)$).

The following family of polynomial kernels have been considered in the literature

$$
K(x, y) = (1 + x \cdot y)^d
$$

for $d > 1$. With $d = 1$, the kernel record linkage corresponds to the distance-based record linkage with the Euclidean distance.

5.6.2   Distance and Normalization

In the previous definitions we have been using $d_i(a, b)$ defined as $d_{V_i}(V_i^A(a), V_i^B(b))$. However, we have not given much detail on how to compute $d_{V_i}(V_i^A(a), V_i^B(b))$. In part, this refers to the expressions in Sect. 5.4.7 where distances at variable level where considered.

At this point we want to underline that record linkage needs to consider standardization of the data. Otherwise variables with large values would weight more than variables that only take values in a small interval. The following two different standardizations have been considered in the literature. The first one corresponds to an attribute-standardizing distance (das).

$$
das_V^2(V^A(a), V^B(b)) = \left(\frac{V^A(a) - \overline{V^A}}{\sigma(V^A)} - \frac{V^B(b) - \overline{V^B}}{\sigma(V^B)} \right). \tag{5.43}
$$

The second one corresponds to a distance-standardizing distance (dds).

$$
dds_V^2(V^A(a), V^B(b)) = \left(\frac{V^A(a) - V^B(b)}{\sigma(V^A - V^B)} \right). \tag{5.44}
$$

Both definitions are for numerical values, but the second one can be easily modified to be appropriate for non numerical, e.g. categorical, values. The modification consists of using the distance between values instead of the difference, and using the σ of the distance instead of the one of the difference.

$$dds_V'^2(V^A(a), V^B(b)) = \left(\frac{d(V^A(a), V^B(b))}{\sigma(d(V^A(a), V^B(b)))} \right)$$

5.6.3 Parameter Determination for Record Linkage

As we have seen, distance-based record linkage requires a distance. Distances depend implicitly or explicitly on weights for the variables (and on the interactions between the variables). For distance based record linkage on numerical data it is common to use the Euclidean distance between pairs of records. We have already discussed briefly alternative distances in the previous section.

In order to have better estimation of the worst-case disclosure risk, we can try to tune the parameters in an optimal way so that the number of reidentifications is maximum. To do so, we can use machine learning and optimization techniques, assuming that we know which are the correct links (i.e., supervised machine learning).

Although this section focuses on supervised approaches for distance based record linkage, we could proceed in a similar way for probabilistic record linkage.

Let us first recall that in Eq. 5.41 we have expressed the distance between pairs of records in terms of the arithmetic mean, and that this expression was later generalized in Definitions 5.14 and 5.15 so that other parametric distances were taken into account. In particular, we have seen that we can use a weighted distance (using the weighted mean, the OWA or the Choquet integral) or a symmetric bilinear form.

In general, given records a and b on variables $\mathbf{V} = \{V_1, V_2, \cdots V_n\}$, we can define a distance between a and b in terms of an aggregation (combination) function $\mathbb{C}$ that aggregates the values of the distances between each variable. If we denote the distance for the ith variable as $d_{V_i}(V_i^A(a), V_i^B(b))$, the distance between a and b corresponds to

$$d^2(a, b) = \mathbb{C}_p(d_{V_1}(V_1^A(a), V_1^B(b)), \ldots, d_{V_n}(V_n^A(a), V_n^B(b)))$$

where p is the parameter of the function $\mathbb{C}$. In Definitions 5.14 and 5.15 we considered different alternatives for $\mathbb{C}$. For the sake of simplicity, we will use $\mathbb{C}_p(a, b)$ to denote the following.

$$\mathbb{C}_p(a, b) = \mathbb{C}_p(d_{V_1}(V_1^A(a), V_1^B(b)), \ldots, d_{V_n}(V_n^A(a), V_n^B(b))).$$

If $\mathbb{C}$ is known, to find the worst-case disclosure risk is to find the parameter p that permits us to achieve the maximum number of reidentifications. When A and B are given and we know which pairs correspond to the same record, this problem can be formulated mathematically. Note that this information is available to those that protect the file. For the sake of simplicity we assume that A and B consist of the same number of records and that records are aligned. That is, a_i and b_i correspond to the same individual.

Using this notation, it is clear that the record a_i is correctly linked to b_i when the following equation holds

$$\mathbb{C}_p(a_i, b_i) < \mathbb{C}_p(a_i, b_j) \tag{5.45}$$

for all $b_j \neq b_i$ in B.

Ideally, we could require all a_i in A to behave in this way, and eventually get the parameter p. This would correspond to have $|A| \times |B|$ equations and find the parameter p that satisfies all of them. Unfortunately, this is usually not possible because there is no p with 100% success. In contrast, we need to permit some incorrect links. In other words, the optimal solution is only able to link correctly some of the pairs. In this case, we can still formalize the problem mathematically. This is done by means of considering blocks of equations.

We consider a block as the set of equations that correspond to a certain record a_i. We then consider a variable K_i associated to this block. When $K = 0$ it means that the constraints are satisfied for a_i, and when $K = 1$ it means that the constraints are not satisfied for a_i. Then, we want to minimize the blocks (i.e., the records a_i) that do not satisfy the constraints. This approach is based on the fact that if there is j_0 such that

$$\mathbb{C}_p(a_i, b_i) > \mathbb{C}_p(a_i, b_{j_0})$$

for $i \neq j_0$ then as the pair (a_i, b_i) is not correctly linked, it does not matter which are the values of $\mathbb{C}_p(a_i, b_i)$ and $\mathbb{C}_p(a_i, b_{j_0})$. I.e., what is relevant is the number of correct and incorrect blocks not the distances themselves.

Using the variables K_i and a constant C that corresponds to the maximum insatisfaction of the inequality, we can rewrite the equation above (Eq. 5.45). We have that for a given record a_i the equation

$$\mathbb{C}_p(a_i, b_j) - \mathbb{C}_p(a_i, b_i) + CK_i > 0.$$

has to be satisfied for all $j \neq i$, for a given C, and for $K_i \in \{0, 1\}$.

Then, the optimization problem is to find the p that minimizes the number of K_i equal to one. Naturally, this is equivalent to minimize $\sum K_i$. This sum will be the number of times we are not able to reidentify correctly, $|B| - \sum K_i$ will be the correct number of reidentifications.

Therefore, the optimization problem results as follows:

Minimize $\sum_{i=1}^{N} K_i$
Subject to:
$$\mathbb{C}_p(a_i, b_j) - \mathbb{C}_p(a_i, b_i) + CK_i > 0, \qquad \forall i = 1, \ldots, |A|,$$
$$\forall j = 1, \ldots, |B|, i \neq j$$
$$K_i \in \{0, 1\}$$

We need to add to this optimization problem constraints on the parameter p if the function $\mathbb{C}$ requires such constraints for the parameter. For example, if we consider $\mathbb{C}$ to be the weighted mean (so, the distance is e.g. a weighted Euclidean distance) we need p to be a weighting vector $p = (p_1, \ldots, p_n)$ with all weights positive $p_i \geq 0$ and adding one $\sum p_i = 1$. This results into the following optimization problem.

$$
\begin{aligned}
&\text{Minimize } \sum_{i=1}^{N} K_i \\
&\text{Subject to:} \\
&\quad \mathrm{WM}_p(a_i, b_j) - \mathrm{WM}_p(a_i, b_i) + C K_i > 0, \qquad \forall i = 1, \ldots, |A|, \\
&\qquad\qquad\qquad\qquad\qquad\qquad\qquad\qquad\qquad \forall j = 1, \ldots, |B|, i \neq j \\
&\qquad\qquad\qquad\qquad\qquad\qquad\qquad\qquad\qquad\qquad\qquad K_i \in \{0, 1\} \\
&\quad \sum_{i=1}^{n} p_i = 1 \\
&\quad p_i \geq 0 \qquad\qquad\qquad\qquad\qquad\qquad\qquad\qquad \text{for } i = 1, \ldots, n
\end{aligned}
$$

$$(5.46)$$

Other distances in Definitions 5.14 and 5.15 imply other constraints on the parameters p. Constraints for the OWA (linear combination of order statistics) are similar (w is a weighting vector with positive elements that add to one). The constraints for the Choquet integral force μ to be a fuzzy measure (i.e., $\mu(\emptyset) = 0$, $\mu(\mathbf{V}) = 1$ and for all $A, B \subseteq \mathbf{V}$ such that $A \subseteq B$ we have that $\mu(A) \leq \mu(B)$), and the constraints for the symmetric bilinear form are to force the matrix Q to be symmetric and semidefinite positive.

The optimization problem described in Eqs. 5.46 with the constraints for the weighted mean, the OWA and the Choquet integral, is a linear optimization problem with linear constraints. As some of the variables are integers (i.e., K_i), it is in fact a Mixed Integer Linear mathematical optimization problem (MILP). This type of problems can be solved effectively using standard solvers. When we use a bilinear form, the problem is more complex because the matrix Q needs to be semidefinite positive, and this cannot be expressed in terms of linear constraints. Semi-definite programming can be used to solve this problem.

References [113–116] have studied these optimization problems for disclosure risk assessment and solved it using IBM ILOG CPLEX tool [117] (version 12.1). Reference [114] uses the weighted mean and the OWA, [113] the weighted mean, and [115] the Choquet Integral. Reference [116] studies the distance based on the bilinear form. In this case, in order to make the optimization problem linear, it replaces the constraints on the matrix Q to be semi-definite positive by a set of equations requiring the distance for the records under consideration to be positive. That is, for all a, b in A and B it requires $SB_Q(a, b) \geq 0$.

It can be proven mathematically that the distance based on the Choquet integral always outperforms the weighted mean and the OWA, and that the bilinear form outperforms the weighted mean. However, using such distances is at the cost of an optimization problem with more constraints, and more difficult to solve. E.g., for $t = |A| = |B|$ records and n variables we need

$$t(t-1) + t + n + 1$$

constraints for the weighted mean (and the OWA) and we determine n parameters, but for the Choquet integral we look for $2^n - 1$ parameters with a problem of

$$ t(t - 1) + t + 2 + \sum_{k=2}^{n} \binom{n}{k} k $$

constraints, and in the case of the bilinear form we look for $n(n - 1)/2$ parameters using $t(t - 1) + t + t^2$ constraints.

Reference [116] compared these distances for 6 different data files and showed that the distance based on the bilinear form was the one that leads to the largest number of reidentifications. This however depends on the files and the type of protection. It seems that when variables are protected independently, the Euclidean distance and weighted Euclidean distance are enough to give a good upper bound of the risk (the improvement achieved is not very much significant). When sets of variables are protected together, other distances as the ones based on the Choquet integral and the bilinear form may be preferable.

In any case, results depend on specific original and protected data files.

5.7   Record Linkage Without Common Variables

When two databases do not share the schema, reidentification may still be possible if the information in both databases is similar. The following hypothesis makes explicit when reidentification may take place.

In order to achieve reidentification in this scenario some assumptions are needed. They are summarized next.

Hypothesis 1 [57] A large set of common individuals is shared by both files.

Hypothesis 2 [57] Data in both files contain, implicitly, similar structural information. In other words, even though there are no common variables, there is some amount of redundancy between variables in both files.

In this second hypothesis structural information corresponds to an explicit representation of the underlying relationship between the individuals in the database. Once this structural information is extracted from the data, reidentification algorithms can be applied to the extracted structure.

This approach was explored in [57] where the structural information was expressed by means of partitions. The rationale is that it is possible to extract from data explicit relationship between individuals and these relationships could be extracted from both files. More particularly, that a clustering algorithm applied to both files would result into similar clusters. If the information in both files is similar, records similar in one file are also similar in the other. This is made explicit in the following hypothesis.

Hypothesis 3 [57] Structural information can be expressed by means of partitions.

Partitions were selected against other data structures as dendrograms, as the former are more robust to changes in the data (this is shown in [118]).

5.8 k-Anonymity and Other Boolean Conditions for Identity Disclosure

k-Anonymity is a Boolean privacy model that focuses on identity disclosure. It was defined by Samarati and Sweeney in [26, 28, 119, 120]. The definition is closely related to the concept of P-set defined by Dalenius in [16] (reviewed in Sect. 5.1), and to quasi-identifiers.

Similar concepts appears in previous works where a minimum and maximum number of records are required to answer a query or to contribute to the cells in a table. For example, focusing on querying databases, [121] discusses an approach that is called Restricted Query Set Size (RQSS) where a query is denied by a database management system if it is found to involve fewer than k or greater than $|X| - k$ records.

Definition 5.17 A data set X satisfies k-anonymity with respect to a set of quasi-identifiers when the projection of X in this set results into a partition of X in sets of at least k indistinguishable records.

Given a database, a set of quasi-identifiers and a value of k, we can certify whether the database is compliant with k-anonymity or not.

For example, the database represented in Table 1.2 (reproduced here as Table 5.8) satisfies k-anonymity for $k = 2$ for the quasi-identifiers (*city, age, illness*).

Note that k-anonymity implies that when we consider any subset of the quasi-identifiers we have an anonymity set of at least k records. Similarly, from the point of view of reidentification, it means that any record linkage algorithm cannot distinguish among the at least k records of the anonymity set (they are equally probable to correspond to a given individual).

Table 5.8 The data in this table permits an adversary to achieve attribute disclosure without identity disclosure. Only attributes City, Age, and Illness are published

Respondent	City	Age	Illness
ABD	Barcelona	30	Cancer
COL	Barcelona	30	Cancer
GHE	Tarragona	60	AIDS
CIO	Tarragona	60	AIDS

5.8.1 *k*-Anonymity and Anonymity Sets: *k*-Confusion

It is important to note that from the reidentification point of view what is relevant in k-anonymity is that the at least k records in the anonymity set are equally probable, or equivalently that the anonymity set of any record has cardinality at least k. It is not so important the fact that all these records have exactly the same values. This observation has led to several extensions of k-anonymity: k-indistinguishability, (k, t)-confusion [122], k-confusion [123], k-concealment [124], probabilistic k-anonymity [125], and crowd blending [126].

Following [123], we define k-confusion as follows.

Definition 5.18 If the cardinality of the anonymity set for any record of a database is at least k, then we say that we have k-confusion.

Note that k-confusion does not establish any requirement on the values of the masked data set. It is clear that k-confusion implies that the probability of reidentification is lower than or equal to $1/k$. The larger the anonymity set, the lower the probability. Probabilistic k-anonymity focuses directly on the probabilities of reidentification. It requires that the probability of reidentification is lower than or equal to $1/k$ for any external dataset. However, probabilistic k-anonymity does not require that all the probabilities are equal. In other words, there is no need that all elements of the anonymity set are equally probable. Because of that, we may have that the probability of one of the elements of the anonymity set is significantly larger than the others. (k, t)-confusion forces that the anonymity set for a given probability t is large enough (at least k).

Definition 5.19 [122] Given a space of databases D, a space of auxiliary information A and a method of anonymization of databases ρ, we say that ρ provides (k, t)-confusion if for all reidentification methods r, and all anonymized databases $X \in D$ the confusion of r with respect to $\rho(X)$ and A is larger or equal to k for the fixed threshold $0 < t \leq 1/k$.

Definition 5.20 [122] Let notations be as in Definition 5.19. We say that an anonymization method ρ provides n-confusion if there is a $t > 0$ such that ρ provides (n, t)-confusion.

These extensions are related to k-concealment. The definition of k-concealment is for the case of using generalization as the masking method for a database. Generalization (see Sect. 6.2.1 for details) is about changing values by more general ones. Let us consider a dataset with records $X = \{x_1, \ldots, x_n\}$, and a masked file $X' = \{x_1', \ldots, x_n'\}$ where each x_i' has been obtained from x_i through generalization. For example, if attribute A_j gives the place of birth for each x_i in the database, then if $A_j(x_i)$ are cities, a possible generalization is to assign $A_j(x_i')$ to the corresponding county. In general, $A_j(x_i) \subseteq A_j(x_i')$. A record x' is said to be consistent with another record x when x' can be seen as a generalization of x.

Definition 5.21 [124, 127] Given a database X and X' a generalization of X; then,

- X' is called a $(1, k)$-anonymization of X if each record in X is consistent with at least k records in X',
- X' is called a $(k, 1)$-anonymization of X if each record in X' is consistent with at least k records in X,
- X' is called a (k, k)-anonymization of X if X' is both a $(1, k)$- and a $(k, 1)$-anonymization of x.

Tassa et al. in [124] discuss these three types of anonymization.

A typical adversarial attack aims at revealing sensitive information on a specific target individual. In such an attack, the adversary knows a record $x \in X$ and he tries to locate the corresponding generalized record $x' \in X'$. Alternatively, the adversary could be interested in reidentifying any entity in the released data, for instance to find possible victims to blackmail. Such an attack works in the opposite direction: Focusing on a generalized record $x' \in X'$, the adversary tries to infer its correct preimage $x \in X$. (...) The notion of $(1, k)$-anonymity aims at protecting against the first attack; the notion of $(k, 1)$-anonymity aims at protecting against the second one; (k, k)-anonymity considers both attacks.

In order to define k-concealment, we need to represent the relationships of consistency between pairs of records (x, x') by means of a bipartite graph. The definition follows.

Definition 5.22 Let X be a database, X' a generalization of X. The bipartite graph $G_{X, X'}$ is defined as the graph with nodes $X \cup X'$ and edges (x, x') if x' is consistent with x.

Recall that in graph theory, we have a perfect matching in a graph bipartite when there is a one-to-one correspondence between the vertexes of the graph. The concept of perfect matching permits to define k-concealment.

Definition 5.23 Let X be a database, X' a generalization of X and $G_{X, X'}$ be the corresponding bipartite graph. We say that $x' \in X'$ is a match of $x \in X$ if (x, x') is an edge of $G_{X, X'}$ and there exists a perfect matching in $G_{X, X'}$ that includes that edge. If all records $x \in X$ have at least k matches in X', then X' is called a k-concealment of X.

Reference [124] points out that this is a generalization of $(1, k)$-anonymity and that similar generalizations can be defined for $(k, 1)$-anonymity and (k, k)-anonymity.

Let us return to our discussion on n-confusion. Note that although the definition of n-confusion focuses only on the anonymity set, reidentification can be seen as explicit. Efficient reidentification algorithms would reduce the cardinality of the anonymity set of any record using as much knowledge and inference as possible. In the case of data protection using generalization, matching and perfect matching would be used to attack a database. In this way, n-confusion and n-concealment can be seen as related, being n-confusion applicable when masking methods other than generalization are used.

We introduce now an example (inspired in [122]) to illustrate the difference between k-anonymity and k-confusion. In particular, we will show that k-confusion permits us to define methods that lead to solutions with a lower information loss than the ones that satisfy k-anonymity.

Example 5.6 Let us consider the following data set X consisting of 4 points in $\mathbb{R}^2$:

$$X = \{(1, 2), (-2, 4), (4, -2), (-3, -3)\}.$$

If we want to protect this data set with k-anonymity with $k = 4$ one way is to replace the four points by their mean. That is,

$$X' = \{(0, 0), (0, 0), (0, 0), (0, 0)\}.$$

In this case, we will have a data set satisfying k-anonymity for $k = 4$ and with the same mean as the original file. However, the deviation of the attributes of the original data set is not equal to the deviation of the attributes of the protected data set. While the corrected sample standard deviation of the first attribute in X is $\sqrt{10}$ and the one of the second attribute in X is $\sqrt{12.8333}$, the deviations of the attributes in the protected data set are zero.

Using k-confusion as our anonymity model we can have four protected records with different values. Let us consider the following protected records

$$X'' = \{(x, 0), (-x, 0), (0, y), (0, -y)\}.$$

Note that these records satisfy the property that their mean is $(0, 0)$ and, thus, satisfy the same property as the data in X'. In addition, we can compute x and y so that the deviations of X'' are equal to the deviations of X. To do so, see that the deviation for the first and the second attribute in X'' are, respectively,

$$s_x = \sqrt{\frac{1}{n-1}\sum(x_i - \bar{x})^2} = \sqrt{\frac{1}{3}(x^2 + x^2 + 0 + 0)} = x\sqrt{2/3},$$

$$s_y = \sqrt{\frac{1}{n-1}\sum(y_i - \bar{x})^2} = \sqrt{\frac{1}{3}(y^2 + y^2 + 0 + 0)} = y\sqrt{2/3}.$$

Therefore, using the equalities $\sqrt{10} = x\sqrt{2/3}$ and $\sqrt{12.83333} = y\sqrt{2/3}$ we have:

Fig. 5.5 Original data set X
(*left*) and protected data set
X'' satisfying k-confusion
(*right*)

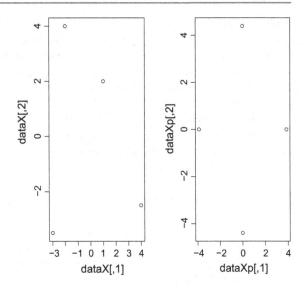

$$x = \sqrt{10}/\sqrt{2/3} = 3.872983$$

$$y = \sqrt{12.8333}/\sqrt{2/3} = 4.387476$$

Figure 5.5 represents the original set X and the protected set X''.

Now, if we publish this data set, and following the transparency principle we explain also how the data set has been produced, we have that a record linkage algorithm exploiting this information will state that the four masked records in X'' are indistinguishable with respect to each of the records in X, and will assign probabilities of reidentification equal to 1/4. So, this dataset has the same (identity) disclosure risk than the k-anonymous X'.

This is a simple example but we can proceed in the same way with larger data sets ensuring k-confusion.

5.8.2 k-Anonymity and Attribute Disclosure: Attacks

It is important to note that k-anonymity does not exclude disclosure. This is in fact a consequence of what was explained in Sect. 1.3.3. We may have attribute disclosure with no identity disclosure.

Disclosure may occur if all the values of a confidential attribute are the same for a particular combination of quasi-identifiers. The following attacks are considered in the literature (see, for example, [128]). These attacks give names to problems already considered in Sects. 1.3.3 and 5.1 (collusion of individuals with respect to a P-set of data).

- **Homogeneity attack**. When all indistinguishable records in a cluster are also indistinguishable with respect to a confidential attribute, disclosure can take place. When the intruder links a record with this cluster, attribute disclosure takes place as the confidential attribute is also known for this record.

- **External knowledge attack**. In this case, some information about an individual is used to deduce information of the same or another individual. Let us consider the case of k-anonymity with $k = 5$, and a cluster cl_0 with 5 records, four of which with illness il_1 and one with illness il_2. Let us assume that our knowledge about the quasi-identifiers of two individuals i_1 and i_2 permits us to deduce that both are in the same cluster cl_0, and that we know that i_1 has no illness il_2, then it is clear that i_1 has illness il_1. Similarly, if we know that i_2 has illness il_2 it is also clear that i_1 has illness il_1.

Some definitions have been introduced in the literature to establish when a data set satisfies certain levels of identity disclosure and attribute disclosure. One of them is p-sensitive k-anonymity, another l-diversity.

Definition 5.24   [129] A data set is said to satisfy p-sensitive k-anonymity for $k > 1$ and $p \leq k$ if it satisfies k-anonymity and, for each group of records with the same combination of values for quasi-identifiers, the number of distinct values for each confidential value is at least p (within the same group).

To satisfy p-sensitive k-anonymity we require p different values in all sets. This requirement might be difficult to achieve in some databases. In others may be easy but at the expenses of a large information loss. A discussion on this model and some of its variations can be found in [130].

Another alternative definition to ensure some level of attribute disclosure is l-diversity [131]. Similar to the case of p-sensitivity, l-diversity forces l different categories in each set. However, in this case, categories should have to be *well-represented*. Different meanings have been given to what *well-represented* means.

Another alternative approach to standard k-anonymity is t-closeness [132]. In this case it is required that the distribution of the attribute in any k-anonymous subset of the database is similar to the one of the full database. Similarity is defined in terms of the distance between the two distributions and such distance should be below a given threshold t.

This approach permits the data protector to limit the disclosure. Nevertheless, a low threshold forces all the sets to have the same distribution as the full data set. This might cause a large information loss: any correlation between the confidential attributes and the one used for l-diversity might be lost when forcing all the sets to have the same distribution.

5.8.3 Other Related Approaches to k-Anonymity

In the above definition for k-anonymity, we require that this property holds for all quasi-identifiers. Reference [133] considers a variation based on a set of *virtual identifiers*. Each virtual identifier consists of a set of attributes and includes an anonymity threshold. Then, for a virtual identifier v_i with anonymity threshold t_i a database satisfies its anonymity requirement when there are at least t_i records in each of the combinations of values for attributes in v_i (zero records are also allowed).

This approach is related to the computational anonymity introduced in [134]. In this work, the author distinguishes between theoretical anonymity and computational anonymity (in a way analogous to theoretical PIR and computational PIR):

> We say that unconditional anonymity is theoretical anonymity. Computational anonymity is conditioned by the assumption that the adversary has some limitation. The limitations can be (…) restricted memory or knowledge.

Computational anonymity can be seen as the underlying scenario for (k, l)-anonymity. We review first the definition of computational anonymity.

Assumption 1 [134] Let X be the table that we want to protect. Then we assume that the information the adversary has about any individual is limited to at most l of the attributes in X. The l attributes do not have to be the same for different individuals.

The definition of (k, l)-anonymity based on computational anonymity follows. The first definition of (k, l)-anonymity was given in [135] in the context of data privacy for graphs.

Definition 5.25 [134] A data set X satisfies (k, l)-anonymity if it is k-anonymous with respect to every subset of attributes of cardinality at most l.

5.9 Discussion on Record Linkage

We finish this chapter with some final remarks related to record linkage and reidentification algorithms. We begin with the issue of formalizing record linkage (why this is needed and how we can formalize it), we follow with a comparison of record linkage algorithms, and we finish on aspects related to record linkage and big data.

5.9.1 Formalization of Reidentification Algorithms

The definition of k-confusion and (k, t)-confusion (Definitions 5.19 and 5.20) and Example 5.6 poses the question of what is a reidentification algorithm. Recall that they are defined in terms of *all reidentification methods*. Naturally, we do not expect that random methods that by chance achieve 100% reidentifications are taken into account. Note that we also want to discard methods that do not take into account information properly.

Let us consider the data of Example 5.6. Four artificial points

$$X'' = \{(x, 0), (-x, 0), (0, y), (0, -y)\}$$

with $x = 4$ and $y = 4$ have been created from four original points

$$X = \{(1, 2), (-2, 4), (4, -2), (-3, -3)\}.$$

There is no direct correspondence between one particular record in X'' and one in X. Nevertheless, the generation of X'' has (*by chance*) assigned the records in this order. Then, $x_1'' = (4, 0)$, $x_2'' = (-4, 0)$, ...

If we apply distance-based record linkage using an Euclidean distance to these pair of records, we will have the following assignment: $x_1 = (1, 2)$ corresponds to $x_2'' = (-4, 0)$, $x_2 = (-2, 4)$ to $x_3'' = (0, 4)$, $x_3 = (4, -2)$ to $x_1'' = (4, 0)$, and $x_4 = (-3, -3)$ to $x_4'' = (0, -4)$. Therefore, we reidentify one records.

Nevertheless, it is wrong to state that the worst-case scenario reidentifies any record because it is just by chance that the records in X'' are ordered in this way. A proper reidentification algorithm would assign a probability of 1/4 to the link with any record or, equivalently, define an anonymity set of four records. Under a proper definition of reidentification algorithm, it is clear that the method of Example 5.6 satisfies $k = 4$-confusion.

References [136, 137] addresses the problem of formulating mathematically what a proper record linkage algorithm is. The definition is based on true probabilities, and it is assumed that due to some uncertainty these probabilities are not known. Then, we can model our uncertainty with imprecise probabilities or belief measures which should be consistent with the true probabilities. Only the algorithms that return probabilities compatible with these belief functions can be called reidentification algorithms. Such definition avoids that the wrong application of the distance-based record linkage above is considered as a reidentification algorithm. Therefore, we can correctly state that the approach in Example 5.6 satisfies $k = 4$-confusion.

This mathematical formulation was applied in [138] to different data masking methods.

5.9.2 Comparison of Record Linkage Algorithms

The comparison of the different generic record linkage algorithms in experiments show that, in general, probabilistic record linkage and distance-based record linkage with unweighted distances lead to similar results for most of the distances used. See, for example, [18] for the case of numerical data and [98] for categorical.

In [27] we showed that the Mahalanobis distance was for some very particular experiments significantly better than the other distances considered (Euclidean, kernel-based distances). For most of the experiments the results were similar to the results of other methods.

The use of supervised approaches to determine the parameters of the distances, as explained in Sect. 5.6.3 has also been proven successful. It is clear that using record linkage with weighted means and other kind of distances that generalize the Euclidean and the Mahalanobis distance should lead to larger reidentification rates than when simply using the Euclidean and Mahalanobis distance. Nevertheless, the computational cost of these approaches can be very large, and even unfeasible to get the optimal solution in reasonable time. Reference [113] describes execution times when finding the parameters of the weighted distance (using the weighted mean).

We have discussed in Sect. 5.3.5 that specific record linkage algorithms are the ones that use information on how data has been protected to improve their performance. Specific record linkage approaches (i.e., transparency attacks) have been proven to be more successful in achieving good reidentification rates. Methods exist to attack files protected with microaggregation [48–50] and rank swapping [30]. We discuss a few transparency attacks in Chap. 6.

Finally, as reported before, distance-based record linkage in its basic version is simpler to implement and operate than probabilistic record linkage. Distance-based record linkage permits us to incorporate easily knowledge on the variables. Optimal parameters for distance-based record linkage are not easy to find if we do not have a training set. When the training set is available, machine learning and optimization techniques can be used for learning the parameters. This is precisely the case in data privacy when risk is estimated by the people that protects the data. They have both the original and the protected files. Nevertheless, the implementation of the algorithms to learn the optimal parameters introduces more complexity to distance-based record linkage.

Both distance-based and probabilistic record linkage have considered the case of non-independent variables. The Mahalanobis distance and the distance-based on the Choquet integral are used in this case within the distance-based record linkage approach.

Two final differences between the two approaches are the following. Distance-based record linkage has been applied taking into account semantic distances between categorical terms appearing in the data, this has not yet incorporated into probabilistic approaches. On the other hand, probabilistic record linkage has been studied taking into account that not all values are equally probable. This issue discussed in Sect. 5.5.4.1 has not yet been considered in distance-based approaches.

5.9.3 Disclosure Risk and Big Data

We can apply the privacy models studied in this chapter to big data. We can use reidentification, k-anonymity, uniqueness, and differential privacy to model disclosure risk.

Nevertheless, actual computations of risk is difficult for big data. One of the reasons is the computational cost. We have discussed above blocking as one of the tools to deal with this problem. Besides of that, in the context of reidentification and k-anonymity for data privacy it is fundamental to establish the set of quasi-identifiers. With big data the amount of information available to intruders can be very large and thus the set of quasi-identifiers is also large. All variables may belong to this set.[4] In addition to the information available, effective machine learning and data mining algorithms can infer the values of (sensitive) variables from other variables available. E.g. there are results inferring sexual orientation, political and religious affiliation, and even whether parents separated before users of social networks were 21 years old [139]. Such inferences can then be used for reidentification. Linking databases, customary in big data, increases the number of variables describing an individual. This makes protection more difficult and increases disclosure risk. It is important in this context to recall what we have stated in Sect. 5.3.5: the worst-case scenario for reidentification is when the intruder uses the original data for attacking a data file. This is also the case for big data.

Another important aspect that needs to be clear is the goal of our protection. It is usual to consider that the goal is to protect individuals and not single records. It is usually implicit that records in a database are independent. However, this is not necessarily so. In fact, disclosure risk and also information loss depend on our assumptions on the data (see e.g. [140]). In any case, large databases may contain several data items corresponding to the same individual even if this is not explicit. Observe the case of anonymizing search logs or shopping carts. Therefore, we need to consider a global perspective for all these records and protect them together. Independent protection of these records may led to disclosure. For example, k-anonymization of k records from the same individual can lead to disclosure. Detecting and solving these situations in big data may be difficult. E.g. a user can use different devices (and accounts) for accessing a search service and, thus, it may be difficult to detect that different logs belong to the same user if there is no identifier linking them.

Releasing several protected copies of data that change over time (i.e., multiple releases as for dynamic data) also pose disclosure problems. Independent protections of the files do not ensure that individuals are protected when considering all data together. Observe Example 4 in [141]. They consider two independently released data sets where both data sets contain the attribute month of birth in a school class. There are two students born in each month except for February. Only one student was born in February. Let us consider that we k-anonymize this file with $k = 2$. We can assign this student to January or to March, resulting into two different $k = 2$-anonymous data sets. The release of these two sets will disclosure that there is a student born in February. Because of that masking needs to be done considering previous releases. See e.g. [141–145] for details on the attacks (and solutions).

---

[4]The discussion of which are the quasi-identifiers for attacking a database is present in e.g. the literature on data protection for graphs and social networks. There are a few competing definitions of k-anonymity for graphs that correspond to different sets of quasi-identifiers. We will discuss them in Sect. 6.4.2 (on algorithms for k-anonymity for big data).

5.9.4 Some Guidelines and Research Issues

As a summary of this section, we give some guidelines related to disclosure risk in big data, and how to decrease this risk. The first two corresponds to the importance of exploiting user privacy and providing technology to increase user awareness of privacy risk. The last ones relate to the application of masking methods to the data and thus are more oriented to respondent and holder privacy. See [146–149]) for additional details on these issues.

- User privacy and decentralized anonymity should be put in place (whenever possible) so that users anonymize their data in origin. In this way there is no need to trust the data collector.
- Technology for user privacy should increase user awareness of privacy risks. That is, technology should help people to know what others know or can infer about them. User interfaces should help users to declare some variables sensitive but also inform what (which other variables, links to friends in social networks, membership to societies, etc.) can be used to infer such sensitive data.
- Databases should be anonymized in origin, before data integration and database linkage, to reduce disclosure risk. Data mining provides algorithms that are resistant to errors, so we can still apply them when data is anonymized. Data utility may be low for those applications in which we focus on the outliers (as e.g. anomaly detection[5]) or, in general, to individual records. In this discussion it is relevant the comment from Little [29] (p. 408) on masking versus information loss:

  > The key distinction is that masking is primarily concerned with identification of *individual* records, whereas statistical analysis is concerned with making inferences about *aggregates*. (...) Methods that exploit this distinction can achieve great gains in confidentiality at little cost. (...) masking becomes inherently difficult when the distinction between aggregate and individual is not clear-cut, as when one large firm dominates a business file, or analysis is required for small subdomains of the population.

- Anonymization needs to provide controlled linkability. So, if data is protected in origin, we can still link them. k-Anonymity permits to link at group level.
- Privacy models need to be composable in order to know the privacy guarantees when protected databases are integrated.

---

[5]There are two main types of methods for anomaly detection: models based on misuses (a database of misuses is used to learn what an anomaly is) and models based on correct activity (the model we learn explains normal activity, and what diverges from the model is classified as an anomaly). The latter approach seems more suitable when data is protected if this process can eliminate outliers, and, thus, the anomalies of a database.

References

1. Bambauer, J.: Tragedy of the deidentified data commons: an appeal for transparency and access. In: Joint UNECE/Eurostat Work Session on Statistical Data Confidentiality, Ottawa, Canada, 28–30 Oct 2013
2. Yakowitz, J.: Tragedy of the data commons. Harv. J. Law Technol. **25**(1), 1–67 (2011)
3. Polonetsky, J., Wolf, C., Brennan, M.W.: Comments of the future of privacy forum. Future of Privacy. http://www.futureofprivacy.org/wp-content/uploads/01-17-2014-FPF-Comments-to-the-FCC.pdf (2014)
4. de Montjoye, Y.-A., Radaelli, L., Singh, V.K., Pentland, A.S.: Unique in the shopping mall: on the reidentifiability of credit card metadata. Science **347**, 536–539 (2015)
5. Jändel, M.: Anonymization of personal data is impossible in practice. Presented in Kistamässan om Samhällssäkerhet (2015)
6. Barth-Jones, D., El Emam, K., Bambauer, J., Cavoukioan, A., Malin, B.: Assessing data intrusion threats. Science **348**, 194–195 (2015)
7. Sánchez, D., Martínez, S., Domingo-Ferrer, J.: Comment on "Unique in the shopping mall: reidentifiability of credit card metadata". Science, 18 March 1274-a (2016)
8. de Montjoye, Y.-A., Pentland, A.S.: Response. Science **348**, 195 (2015)
9. de Montjoye, Y.-A., Pentland, A.S.: Response to Comment on "Unique in the shopping mall: On the reidentifiability of credit card metadata". Science, 18 March 1274-b (2016)
10. Cavoukian, A., El Emam, K.: Dispelling the Myths Surrounding De-identification: Anonymization Remains a Strong Tool for Protecting Privacy (2011)
11. Dalenius, T.: Towards a methodology for statistical disclosure control. Statistisk Tidskrift **5**, 429–444 (1977)
12. Krantz, D.H., Luce, R.D., Suppes, P., Tversky, A.: Foundations of Measurement: Additive and Polynomial Representations, vol. 1. Academic Press, New York (1971)
13. Luce, R.D., Krantz, D.H., Suppes, P., Tversky, A.: Foundations of Measurement: Representation, Axiomatization, and Invariance, vol. 3. Academic Press, New York (1990)
14. Roberts, F.S.: Measurement Theory. Addison-Wesley, Reading (1979)
15. Suppes, P., Krantz, D.H., Luce, R.D., Tversky, A.: Foundations of Measurement: Geometrical, Threshold, and Probability Representations, vol. 2. Academic Press, San Diego (1989)
16. Dalenius, T.: Finding a needle in a haystack—or identifying anonymous census records. J. Off. Stat. **2**(3), 329–336 (1986)
17. Domingo-Ferrer, J., Mateo-Sanz, J.M., Torra, V.: Comparing SDC methods for microdata on the basis of information loss and disclosure risk. In: Pre-proceedings of ETK-NTTS 2001, vol. 2, pp. 807–826. Eurostat (2001)
18. Domingo-Ferrer, J., Torra, V.: A quantitative comparison of disclosure control methods for microdata. In: Doyle, P., Lane, J.I., Theeuwes, J.J.M., Zayatz, L. (eds.) Confidentiality, Disclosure and Data Access: Theory and Practical Applications for Statistical Agencies, pp. 111–134. North-Holland, Amsterdam (2001)
19. Mateo-Sanz, J.M., Sebé, F., Domingo-Ferrer, J.: Outlier protection in continuous microdata masking. In: PSD 2004. LNCS, vol. 3050, pp. 201–215 (2004)
20. Templ, M.: Statistical disclosure control for microdata using the R-Package sdcMicro. Trans. Data Priv. **1**, 67–85 (2008)
21. Nin, J., Herranz, J., Torra, V.: Using classification methods to evaluate attribute disclosure risk. In: Proceedings of the MDAI 2010. LNCS, vol. 6408, pp. 277–286 (2010)
22. Herranz, J., Matwin, S., Nin, J., Torra, V.: Classifying data from protected statistical datasets. Comput. Secur. **29**, 875–890 (2010)
23. Hall, M., Frank, E., Holmes, G., Pfahringer, G., Reutemann, P., Witten, I.H.: The WEKA data mining software: an update. SIGKDD Explor. **11**, 1 (2009)

24. Balsa, E., Troncoso, C., Díaz, C.: A metric to evaluate interaction obfuscation in online social networks. Int. J. Uncertain. Fuzziness Knowl. Based Syst. **20**, 877–892 (2012)
25. Muralidhar, M., Sarathy, R.: Statistical dependence as the basis for a privacy measure for microdata release. Int. J. Uncertain. Fuzziness Knowl. Based Syst. **20**, 893–906 (2012)
26. Sweeney, L.: k-anonymity: a model for protecting privacy. Int. J. Uncertain. Fuzziness Knowl. Based Syst. **10**(5), 557–570 (2002)
27. Torra, V., Abowd, J.M., Domingo-Ferrer, J.: Using mahalanobis distance-based record linkage for disclosure risk assessment. LNCS, vol. 4302, pp. 233–242 (2006)
28. Samarati, P.: Protecting respondents' identities in microdata release. IEEE Trans. Knowl. Data Eng. **13**(6), 1010–1027 (2001)
29. Little, R.J.A.: Statistical analysis of masked data. J. Off. Stat. **9**(2), 407–426 (1993)
30. Nin, J., Herranz, J., Torra, V.: Rethinking rank swapping to decrease disclosure risk. Data Knowl. Eng. **64**(1), 346–364 (2007)
31. Torra, V.: OWA operators in data modeling and reidentification. IEEE Trans. Fuzzy Syst. **12**(5), 652–660 (2004)
32. Domingo-Ferrer, J., Torra, V.: Disclosure risk assessment in statistical microdata protection via advanced record linkage. Stat. Comput. **13**, 343–354 (2003)
33. Spruill, N.L.: The confidentiality and analytic usefulness of masked business microdata. In: Proceedings of the Section on Survey Research Methods 1983, pp. 602–610, American Statistical Association (1983)
34. Paass, G.: Disclosure risk and disclosure avoidance for microdata. J. Bus. Econ. Stat. **6**, 487–500 (1985)
35. Paass, G., Wauschkuhn, U.: Datenzugang, Datenschutz und Anonymisierung—Analysepotential und Identifizierbarkeit von Anonymisierten Individualdaten. Oldenbourg Verlag (1985)
36. Willenborg, L., de Waal, T.: Elements of Statistical Disclosure Control. Lecture Notes in Statistics. Springer, New York (2001)
37. Elliot, M.J., Manning, A.M., Ford, R.W.: A computational algorithm for handling the special uniques problem. Int. J. Uncertain. Fuzziness Knowl. Based Syst. **10**(5), 493–509 (2002)
38. Manning, A.M., Haglin, D.J., Keaner, J.A.: A recursive search algorithm for statistical disclosure assessment. Data Min. Knowl. Disc. **16**, 165–196 (2008)
39. Elliot, M.J., Skinner, C.J., Dale, A.: Special uniqueness, random uniques and sticky populations: some counterintuitive effects of geographical detail on disclosure risk. Res. Off. Stat. **1**(2), 53–67 (1998)
40. Elamir, E.A.H.: Analysis of re-identification risk based on log-linear models. In: Proceedings of the PSD 2004. LNCS, vol. 3050, pp. 273–281 (2004)
41. Elliot, M.: Integrating file and record level disclosure risk assessment. In: Domingo-Ferrer, J. (ed.) Inference Control in Statistical Databases. LNCS, vol. 2316, pp. 126–134 (2002)
42. Franconi, L., Polettini, S.: Individual risk estimation in μ-argus: a review. In: PSD 2004. LNCS, vol. 3050, pp. 262–272 (2004)
43. Winkler, W.E.: Re-identification methods for masked microdata. In: Proceedings of PSD 2004. LNCS, vol. 3050, pp. 216–230 (2004)
44. Bacher, J., Brand, R., Bender, S.: Re-identifying register data by survey data using cluster analysis: an empirical study. Int. J. Uncertain. Fuzziness Knowl. Based Syst. **10**(5), 589–607 (2002)
45. Herranz, J., Nin, J., Rodríguez, P., Tassa, T.: Revisiting distance-based record linkage for privacy-preserving release of statistical datasets. Data Knowl. Eng. **100**, 78–93 (2015)
46. Lenz, R.: A graph theoretical approach to record linkage. In: Joint ECE/Eurostat Work Session on Statistical Data Confidentiality, Working Paper no. 35 (2003)

47. Scannapieco, M., Cibella, N., Tosco, L., Tuoto, T., Valentino, L., Fortini, M., Mancini, L.: Relais (REcord Linkage At IStat): user's guide. http://www.istat.it/en/tools/methods-and-it-tools/processing-tools/relais (2015)
48. Torra, V., Miyamoto, S.: Evaluating fuzzy clustering algorithms for microdata protection. In: Proceedings of PSD 2004. LNCS, vol. 3050, pp. 175–186 (2004)
49. Nin, J., Torra, V.: Analysis of the univariate microaggregation disclosure risk. New Gener. Comput. **27**, 177–194 (2009)
50. Nin, J., Herranz, J., Torra, V.: On the disclosure risk of multivariate microaggregation. Data Knowl. Eng. **67**(3), 399–412 (2008)
51. Narayanan, A., Shmatikov, V.: Robust de-anonymization of large sparse datasets. In: Proceedings of the 2008 IEEE Symposium on Security and Privacy (SP 2008), pp. 111–125 (2008)
52. Martínez, S., Valls, A., Sánchez, D.: An ontology-based record linkage method for textual microdata. In: Artificial Intelligence Research and Development, vol. 232, pp. 130-139. IOS Press (2011)
53. Han, J., Kamber, M., Pei, J.: Data Mining: Concepts and Techniques, 3rd edn. Morgan Kaufmann Publishers, San Francisco (2011)
54. Rahm, E., Do, H.-H.: Data cleaning: problems and current approaches. Bull. IEEE Comput. Soci. Techn. Committee Data Eng. **23**(4), 3–13 (2000)
55. International Classification of Diseases (ICD), 10-th revision. http://www.who.int/classifications/icd/en/. Accessed Jan 2017
56. Gaines, B.R., Shaw, M.L.G.: Knowledge acquisition tools based on personal construct psychology. Knowl. Eng. Rev. **8**, 49–85 (1993)
57. Torra, V.: Towards the re-identification of individuals in data files with non-common variables. In: Proceedings of ECAI 2000, pp. 326–330 (2000)
58. Boose, J.H.: Expertise Transfer for Expert System Design. Elsevier, New York (1986)
59. Christen, P.: Data Matching—Concepts and Techniques for Record Linkage, Entity Resolution, and Duplicate Detection. Springer, Heidelberg (2012)
60. Torra, V., Narukawa, Y.: Modeling decisions: information fusion and aggregation operators. Springer, Heidelberg (2007)
61. Euzenat, J., Shvaiko, P.: Ontology Matching. Springer, Heidelberg (2013)
62. Shvaiko, P., Euzenat, J.: Ontology matching: state of the art and future challenges. IEEE Trans. Knowl. Data Eng. **25**(1), 158–176 (2013)
63. Bergamaschi, S., Castano, S., Vincini, M., Beneventano, D.: Semantic integration of heterogeneous information sources. Data Knowl. Eng. **36**(3), 215–249 (2001)
64. Do, H.-H., Rahm, E.: COMA—a system for flexible combination of schema matching approaches. In: Proceedings of VLDB, pp. 610-621 (2002)
65. Embley, D.W., Jackman, D., Xu, L.: Multifaceted Exploitation of Metadata for Attribute Match Discovery in Information Integration (2001)
66. Princeton University: "About WordNet". WordNet. Princeton University. http://wordnet.princeton.edu (2010). Accessed Jan 2017
67. Rahm, E., Bernstein, P.A.: A survey of approaches to automatic schema matching. VLDB J. **10**, 334–350 (2001)
68. Haas, L.M., Miller, R.J., Niswonger, B., Tork Roth, M., Schwarz, P.M., Wimmers, E.L.: Transforming heterogeneous data with database middleware: beyond integration. Bull. IEEE Comput. Soci. Techn. Committee Data Eng. **22**, 31–36 (1999)
69. Borkar, V., Deshmukh, K., Sarawagi, S.: Automatic segmentation of text into structured records. In: Proceedings of ACM SIGMOD Conference (2001)
70. Churches, T., Christen, P., Lim, K., Zhu, J.X.: Preparation of name and address data for record linkage using hidden Markov models. BMC Med. Inform. Decis. Making **2**, 9 (2002)
71. Li, W.-S., Clifton, C.: SEMINT: a tool for identifying attribute correspondences in heterogeneous databases using neural networks. Data Knowl. Eng. **33**, 49–84 (2000)

72. Winkler, W.E.: Matching and record linkage. Statistical Research Division, U.S. Bureau of the Census (USA), RR93/08 (1993). Also in B.G. Cox (ed.) Business Survey Methods, pp. 355–384. Wiley (1995)
73. Steorts, R.C., Ventura, S.L., Sadinle, M., Fienberg, S.E.: A comparison of blocking methods for record linkage. In: PSD 2014. LNCS, vol. 8744, pp. 253–268 (2014)
74. Aho, A.V., Ullman, J.D., Hopcroft, J.E.: Data Structures and Algorithms. Addison-Wesley, Reading (1988)
75. Christen, P.: A survey of indexing techniques for scalable record linkage and deduplication. IEEE Trans. Knowl. Data Eng. **24**, 1537–1555 (2012)
76. Nin, J., Muntés-Mulero, V., Martínez-Bazan, N., Larriba-Pey, J.-L.: On the use of semantic blocking techniques for data cleansing and integration. In: Proceedings of IDEAS 2007, pp. 190–198 (2007)
77. Michelson, M., Knoblock, C.A.: Learning blocking schemes for record linkage. In: AAAI (2006)
78. Searcóid, M.O.: Metric Spaces. Springer, London (2007)
79. Salari, M., Jalili, S., Mortazavi, R.: TBM, a transformation based method for microaggregation of large volume mixed data. Data Min. Knowl. Disc. (2016, in press). doi:10.1007/s10618-016-0457-y
80. Navarro, G.: A guided tour to approximate string matching. ACM Comput. Surv. **33**(1), 31–88 (2001)
81. Stephen, G.A.: String Searching Algorithms. World Scientific Publishing Co., Singapore (1994)
82. Odell, M.K., Russell, R.C.: U. S. Patents 1261167 (1918)
83. Odell, M.K., Russell, R.C.: U. S. Patents 1435663 (1922)
84. Knuth, D.E.: The Art of Computer Programming: Sorting and Searching, vol. 3. Addison-Wesley, Reading (1973)
85. Jaro, M.A.: Advances in record-linkage methodology as applied to matching the 1985 Census of Tampa, Florida. J. Am. Stat. Assoc. **84**(406), 414–420 (1989)
86. Newcombe, H.B.: Record linking: the design of efficient systems for linking records into individuals and family histories. Am. J. Hum. Genet. **19 Part I**(3) (1967)
87. Blair, C.R.: A program for correcting spelling errors. Inf. Control **3**(1), 60–67 (1960)
88. Jaro, M.A.: UNIMATCH: a record linkage system: user's manual. U.S, Bureau of the Census, Washington DC (1978)
89. Porter, E.H., Winkler, W.E.: Approximate string comparison and its effect on an advanced record linkage system. Report RR97/02, Statistical Research Division, U.S. Bureau of the Census, USA (1997)
90. This link is currently outdated. http://www.census.gov/geo/msb/stand/strcmp.c. Code, http://www.perlmonks.org/?node=659795, https://people.rit.edu/rmb5229/320/project3/jaro_winkler.html
91. Levenshtein, V.I.: Binary codes capable of correcting deletions, insertions, and reversals, Doklady Academii nauk SSSR **163**(4), 845–848 (1965) (in Russian). (Also in Cybern. Control Theor. **10**(8), 707–710 (1966))
92. Cohen, W.W., Ravikumar, P., Fienberg, S.E.: A comparison of string metrics for matching names and records. In: Proceedings of the KDD 2003 (2003)
93. Herranz, J., Nin, J., Solé, M.: Optimal Symbol Alignment distance: a new distance for sequences of symbols. IEEE Trans. Knowl. Data Eng. **23**(10), 1541–1554 (2011)
94. Muralidhar, K., Domingo-Ferrer, J.: Rank-based record linkage for re-identification risk assessment. In: Proceedings of PSD 2016 (2016)
95. Torra, V., Domingo-Ferrer, J.: Record linkage methods for multidatabase data mining. In: Torra, V. (ed.) Information Fusion in Data Mining, pp. 101–132. Springer, Heidelberg (2003)

96. Fellegi, I.P., Sunter, A.B.: A theory for record linkage. J. Am. Stat. Assoc. **64**(328), 1183–1210 (1969)
97. Dempster, A.P., Laird, N.M., Rubin, D.B.: Maximum likelihood from incomplete data via the EM algorithm. J. Roy. Stat. Soc. **39**, 1–38 (1977)
98. Domingo-Ferrer, J., Torra, V.: Validating distance-based record linkage with probabilistic record linkage. LNCS, vol. 2504, pp. 207–215 (2002)
99. Newcombe, H.B., Kennedy, J.M., Axford, S.L., James, A.P.: Automatic linkage of vital records. Science **130**, 954 (1959)
100. Winkler, W.E.: Methods for Record Linkage and Bayesian Networks, Bureau of the Census (USA), RR2002/05 (2002)
101. Larsen, M.D., Rubin, D.B.: Iterative automated record linkage using mixture models. J. Am. Stat. Assoc. **79**, 32–41 (2001)
102. Winkler, W.E.: Improved decision rules in the Fellegi-Sunter model of record linkage, pp. 274–279. In: Proceedings of the Section on Survey Research Methods. American Statistical Association (1993)
103. Tromp, M., Méray, N., Ravelli, A.C.J., Reitsma, J.B., Bonsel, G.J.: Ignoring dependency between linking variables and its impact on the outcome of probabilistic record linkage studies. J. Am. Med. Inform. Assoc. **15**(5), 654–660 (2008)
104. Daggy, J.K., Xu, H., Hui, S.L., Gamache, R.E., Grannis, S.J.: A practical approach for incorporating dependence among fields in probabilistic record linkage. BMC Med. Inform. Decis. Making **13**, 97 (2013)
105. Herzog, T.N., Scheuren, F.J., Winkler, W.E.: Data Quality and Record Linkage Techniques. Springer, New York (2007)
106. Elmagarmid, A.K., Ipeirotis, P.G., Verykios, V.S.: Duplicate record detection: a survey. IEEE Trans. Knowl. Data Eng. **19**(1), 1–16 (2007)
107. https://www.cs.cmu.edu/~wcohen/matching/. Accessed Jan 2017
108. Winkler, W.E.: Overview of record linkage and current research directions, U.S. Census Bureau RR2006/02 (2006)
109. Vatsalan, D., Christen, P., Verykios, V.S.: A taxonomy of privacy-preserving record linkage techniques. Inf. Syst. **38**, 946–969 (2013)
110. Pagliuca, D., Seri, G.: Some results of individual ranking method on the system of enterprise accounts annual survey. Esprit SDC Project, Deliverable MI-3/D2 (1999)
111. Beliakov, G., Pradera, A., Calvo, T.: Aggregation Functions: A Guide for Practitioners. Springer, Heidelberg (2007)
112. Grabisch, M., Marichal, J.-L., Mesiar, R., Pap, E.: Aggregation Functions. In: Encyclopedia of Mathematics and its Applications, No. 127. Cambridge University Press (2009)
113. Abril, D., Navarro-Arribas, G., Torra, V.: Improving record linkage with supervised learning for disclosure risk assessment. Inf. Fusion **13**(4), 274–284 (2012)
114. Torra, V., Navarro-Arribas, G., Abril, D.: Supervised learning for record linkage through weighted means and OWA operators. Control Cybern. **39**(4), 1011–1026 (2010)
115. Abril, D., Navarro-Arribas, G., Torra, V.: Choquet Integral for Record Linkage. Ann. Oper. Res. **195**, 97–110 (2012)
116. Abril, D., Navarro-Arribas, G., Torra, V.: Supervised learning using a symmetric bilinear form for record linkage. Inf. Fusion **26**, 144–153 (2016)
117. IBM ILOG CPLEX: High-performance mathematical programming engine. International business machines corp. http://www-01.ibm.com/software/integration/optimization/cplex/ (2010)
118. Neumann, D.A., Norton Jr., V.T.: Clustering and isolation in the consensus problem for partitions. J. Classif. **3**, 281–297 (1986)

119. Samarati, P., Sweeney, L.: Protecting privacy when disclosing information: k-anonymity and its enforcement through generalization and suppression. Technical report, SRI International (1998)
120. Sweeney, L.: Achieving k-anonymity privacy protection using generalization and suppression. Int. J. Uncertain. Fuzziness Knowl. Based Syst. **10**(5), 571–588 (2002)
121. Tendick, P., Matloff, N.: A modified random perturbation method for database security. ACM Trans. Database Syst. **19**, 47–63 (1994)
122. Stokes, K., Torra, V.: n-Confusion: a generalization of k-anonymity. In: Proceedings of the Fifth International Workshop on Privacy and Anonymity on Information Society (PAIS 2012) (2012)
123. Stokes, K., Farràs, O.: Linear spaces and transversal designs: k-anonymous combinatorial configurations for anonymous database search. Des. Codes Cryptogr. **71**, 503–524 (2014)
124. Tassa, T., Mazza, A., Gionis, A.: k-Concealment: an alternative model of k-Type anonymity. Trans. Data Priv. **5**(1), 189–222 (2012)
125. Soria-Comas, J., Domingo-Ferrer, J.: Probabilistic k-anonymity through microaggregation and data swapping. FUZZ-IEEE **2012**, 1–8 (2012)
126. Gehrke, J., Hay, M., Lui, E., Pass, R.: Crowd-blending privacy. In: 32nd International Cryptology Conference (CRYPTO 2012) (2012)
127. Gionis, A., Mazza, A., Tassa, T.: k-anonymization revisited. In: Proceedings of ICDE 2008 (2008)
128. Capitani, D., di Vimercati, S., Foresti, S., Livraga, G., Samarati, P.: Data privacy: definitions and techniques. Int. J. Uncertain. Fuzziness Knowl. Based Syst. **20**(6), 793–817 (2012)
129. Truta, T.M., Vinay, B.: Privacy protection: p-sensitive k-anonymity property. In: Proceedings of the 2nd International Workshop on Privacy Data management (PDM 2006), p. 94 (2006)
130. Truta, T.M., Campan, A., Sun, X.: an overview of p-sensitive k-anonymity models for microdata anonymization. Int. J. Uncertain. Fuzziness Knowl. Based Syst. **20**(6), 819–838 (2012)
131. Machanavajjhala, A., Gehrke, J., Kiefer, D., Venkitasubramanian, M.: L-diversity: privacy beyond k-anonymity. In: Proceedings of the IEEE ICDE (2006)
132. Li, N., Li, T., Venkatasubramanian, S.: T-closeness: privacy beyond k-anonymity and l-diversity. In: Proceedings of the IEEE ICDE 2007 (2007)
133. Fung, B.C.M., Wang, K., Yu, P.S.: Top-down specialization for information and privacy preservation. In: Proceedings of ICDR 2005 (2005)
134. Stokes, K.: On computational anonymity. In: Proceedings of PSD 2012. LNCS, vol. 7556, pp. 336–347 (2012)
135. Stokes, K., Torra, V.: Reidentification and k-anonymity: a model for disclosure risk in graphs. Soft Comput. **16**(10), 1657–1670 (2012)
136. Torra, V.: Towards the formalization of re-identification for some data masking methods. In: Proceedings of CCIA 2012, pp. 47–55 (2012)
137. Torra, V., Stokes, K.: A formalization of re-identification in terms of compatible probabilities, CoRR abs/1301.5022 (2013)
138. Torra, V., Stokes, K.: A formalization of record linkage and its application to data protection. Int. J. Uncertain. Fuzziness Knowl. Based Syst. **20**, 907–919 (2012)
139. Kosinski, M., Stillwell, D., Graepel, T.: Private traits and attributes are predictable from digital records of human behavior. PNAS **110**, 5802–5805 (2013)
140. Kifer, D., Machanavajjhala, A.: No free lunch in data privacy. In: Proceedings of SIGMOD 2011 (2011)
141. Stokes, K., Torra, V.: Multiple releases of k-anonymous data sets and k-anonymous relational databases. Int. J. Uncertain. Fuzziness Knowl. Based Syst. **20**(6), 839–854 (2012)
142. Pei, J., Xu, J., Wang, Z., Wang, W., Wang, K.: Maintaining k-anonymity against incremental updates. In: Proceedings of SSDBM (2007)

143. Truta, T.M., Campan, A.: K-anonymization incremental maintenance and optimization techniques. In: Proceedings of ACM SAC 2007, pp. 380–387 (2007)
144. Nergiz, M.E., Clifton, C., Nergiz, A.E.: Multirelational k-anonymity. IEEE Trans. Knowl. Data Eng. **21**(8), 1104–1117 (2009)
145. Navarro-Arribas, G., Abril, D., Torra, V.: Dynamic anonymous index for confidential data. In: Proceedings of 8th DPM and SETOP, pp. 362-368 (2013)
146. D'Acquisto, G., Domingo-Ferrer, J., Kikiras, P., Torra, V., de Montjoye, Y.-A., Bourka, A.: Privacy by design in big data: An overview of privacy enhancing technologies in the era of big data analytics. ENISA Report (2015)
147. Estivill-Castro, V., Nettleton, D.F.: Privacy Tips: Would it be ever possible to empower online social-network users to control the confidentiality of their data? In: Proceedings of ASONAM 2015, pp. 1449–1456 (2015)
148. Soria-Comas, J., Domingo-Ferrer, J.: Big data privacy: challenges to privacy principles and models. Data Sci. Eng. **1**(1), 21–28 (2016)
149. Torra, V., Navarro-Arribas, G.: Big data privacy and anonymization. In: Lehmann, A., Whitehouse, D., Fischer-Hübner, S., Fritsch, L.: Privacy and identity management—facing up to next steps. Springer (2017, in press)

Masking Methods

6

> but when you have to turn into a chrysalis—you
> will some day, you know—and then after that into
> a butterfly
>
> L. Carrol, Alice's adventures in wonderland,
> 1865, Chapter V [1]

In this chapter we review methods that are applicable to data files where each record corresponds to an individual or entity. In the SDC community this is known as microdata. Methods for aggregates of data, what is known as tabular data in the SDC community, have been briefly discussed in Sect. 3.6.

Methods can be classified into three different categories depending on how they manipulate the original data in order to build the protected data set.

- **Perturbative**. The original data set is distorted in some way, and the new data set might contain some erroneous information. E.g. noise is added to an attribute following a normal distribution $N(0, a)$ for a given a. In this way, some combinations of values disappear, and, new combinations appear in the protected data set. As combinations in the masked data set no longer correspond to the ones in the original data set. This obfuscation makes disclosure difficult for intruders.
- **Non-perturbative**. Protection is achieved through replacing an original value by another one that is not incorrect but less specific. For example, we replace a real number by an interval. In general, non-perturbative methods reduce the level of detail of the data set. This detail reduction causes different records to have the same combinations of values, which makes disclosure difficult to intruders.

© Springer International Publishing AG 2017
V. Torra, *Data Privacy: Foundations, New Developments
and the Big Data Challenge*, Studies in Big Data 28,
DOI 10.1007/978-3-319-57358-8_6

- **Synthetic Data Generators**. In this case, instead of distorting the original data, new artificial data is generated and used to substitute the original values. Formally, synthetic data generators build a data model from the original data set and, subsequently, a new (protected) data set is randomly generated using the model. As the published data is artificially created, disclosure risk decreases.

An alternative dimension to classify protection methods is based on the type of data. We can distinguish between numerical and categorical data, but other types of data as e.g. time series, sequences of locations and events, search and access logs, graphs and data from online social networks have also been considered in the literature. Some of the methods we review can be applied to different types of databases.

- **Numerical**. As usual, an attribute is numerical when it is represented by numbers and arithmetic operations. I.e., when addition and substraction can be used in a meaningful way and they lead to sound results. Income and age are typical examples of such attributes. With respect to disclosure risk, numerical values are likely to be unique in a database and, therefore, leading to disclosure if no action is taken.
- **Categorical**. In this case, the attribute takes values over a finite set. Even in the case that a numerical representation is used standard numerical operations do not make sense. Ordinal and nominal scales are typically distinguished among categorical attributes. In ordinal scales the order between values is relevant (e.g. academic degree), whereas in nominal scales it is not (e.g. hair color). Therefore, max and min operations are meaningful in ordinal scales but not on nominal scales.
  Structured attributes is a subclass of categorical attributes. In this case, different categories are related in terms of subclasses or *member of* relationships. In some cases, a hierarchy between categories can be inferred from these relationships. Cities, counties, and provinces are typical examples of these hierarchical attributes. For some attributes, the hierarchy is given but for others it is constructed by the protection procedures.
- **Longitudinal data and time series**. They correspond to sequences of data collected at different times. In longitudinal data, data are obtained from the same sample in all the period of study. This is not usually so usual in time series. Another difference is that time series usually have more repeated observations than studied cases. In contrast, longitudinal studies have more cases than observations. In addition, in time series, uniform time intervals are frequently used. This is not so often the case for longitudinal studies. The stock market index and the unemployment rate are examples of time series. Clinical trials are examples of longitudinal studies.
- **Location data**. This corresponds to sequences of positions and timestamps. It typically corresponds to data gathered from mobile devices. From the perspective of data privacy, it is important to underline that only a few points of a trajectory can lead to disclosure (e.g., home and workplace).

- **Graphs and social networks**. Social networks can be represented as graphs and as such they can be protected. Graphs can include additional information attached to nodes and edges (i.e., they are labeled graphs).
- **Logs**. We may have search and access logs. Internet services keep track of all actions from users. Access logs include additional information as e.g. IP addresses and time. It is usual to have multiple logs from the same person.

We review next some of the existing protection methods following the classification above: Perturbative methods are described in Sect. 6.1, non-perturbative methods in Sect. 6.2, and synthetic data generators in Sect. 6.3. For additional information, reviews on data protection procedures can be found in [2–6].

In addition, we have devoted a section to k-anonymity (Sect. 6.4). As we have discussed already in Sect. 5.8, k-anonymity is not a protection method but a Boolean definition to deal with disclosure risk. We will discuss in Sect. 6.4 methods for obtaining data compliant with this definition. We will also discuss how to protect data when the data to be masked satisfy some constraints (properties) that the masked data also need to satisfy. This discussion is in Sect. 6.5.

In this chapter we will use X to denote the original data, X' to denote the masked data set, and $x_{i,V}$ to represent the value of the attribute V in the ith record. X' is obtained from X applying a masking method ρ to X. That is, $X' = \rho(X)$.

6.1 Perturbative Methods

In this section we review some of the perturbative methods. Among them, the ones that are most used by the statistical agencies are rank swapping and microaggregation [7], but the literature on privacy preserving data mining, more oriented to business-related applications, largely focus on additive noise and microaggregation. Most of the methods described in this section, with some of their variants are implemented in the sdcMicro package in R [8] and in the μ-Argus software [9].

6.1.1 Data and Rank Swapping

Dalenius and Reiss [10,11] introduced data swapping in 1978. The goal of the anonymization process is to preserve t-order frequency tables (contingency tables up to dimension t). The justification is that "for categorical databases, frequency counts embody most of the relevant statistical information".[1] The definition and justification of the method focuses on ordinal data but it can also be applied to numerical data.

---

[1]Note that in Sect. 7.3.2, we review some information loss measures for categorical data based on contingency tables.

Algorithm 18: Random assignment: $Choose(p)$.

Data: p: probability in [0,1]
Result: Boolean value in {0,1}
1 r:= a random value ;
2 **if** $r < p$ **then**
3 $\quad$ **return** 0 ;
4 **else**
5 $\quad$ **return** 1 ;

Let F_t^X denote the t-order frequency table of data set X. That is, F_0^X is the number of records in the data set X, F_1^X is the frequency of a given value of a variable, and F_2^X is the frequency given two values of two given variables. E.g., $F_1^X[V_i = a]$ is the number of records in which the variable V_i is a in the file X, and $F_2^X[V_i = a, V_j = b]$ is the number of records in which the variable V_i is a and variable V_j is b.

Then, data swapping in [10, 11] is defined in terms of swaps between values of one or more variables "subject to the invariance condition that there be" e.g. two-order equivalence. That is, F_2^X equals to $F_2^{X'}$ where X is the original file and X' the masked one.

Reiss [12, 13] proposes an algorithm for Boolean data that gives an approximate solution of the problem for 2-order frequency tables. We reproduce here in Algorithm 19 how to construct the masked data set from the frequency tables of the original data up to dimension two.

The algorithm uses a function called *Choose*, which assigns a Boolean value given a certain probability. See Algorithm 18.

It can be seen that the algorithm generates as much records as in the original file. Then, each record is constructed iteratively with the value of the jth variable assigned taking into account the values already assigned to variables $1, \ldots, j - 1$. In particular, the value for the jth value is zero with a probability defined as the average of the probability of having zero in the original file when the variables $1, \ldots, j - 1$

Algorithm 19: Data swapping approximating second order frequency tables.

Data: F_t: the set of all frequency tables for $0 \leq t \leq 2$, N: number of records in the original file, V number of variables in the original file
Result: X': masked data set consistent with F_t
1 **for** $i = 1$ **to** N **do**
2 $\quad p_1 := F_1[V_1 = 0]/F_0$
3 $\quad X'[i, 1] := Choose(p_1)$
4 $\quad$ **for** $j = 2$ **to** V **do**
5 $\quad\quad p_j := \frac{1}{j-1} \cdot \sum_{k=1}^{j-1} \frac{F_2[V_k=X'_{ij}, V_j=0]}{F_1[V_k=X'_{ik}]}$
6 $\quad\quad X'[i, j] := Choose(p_j)$

have the same values as the ones of this new record. That is,

$$p_j = \frac{1}{j-1} \cdot \sum_{k=1}^{j-1} \frac{F_2[V_k = X'_{ij}, V_j = 0]}{F_1[V_k = X'_{ik}]}$$

Note that $F_2[V_k = X'_{ij}, V_j = 0]/F_1[V_k = X'_{ik}]$ is the estimated probability in the original file that V_j is zero when $V_k = X'_{ij}$. When $F_1[V_k = X'_{ik}]$ we define $p_j = 1/2$.

Reiss proves some properties of their method. In particular, that first-order frequency tables tend to be preserved by any method based on the preservation of second—or third-order frequency tables; and that in the rare case of mutually independent variables, a method that preserves third-order frequency tables will tend to preserve second-order ones. Nevertheless, the author also proves that an algorithm that preserves the first-order frequency tables can lead to data with 50% of error in the second-order ones, and an algorithm preserving the second-order frequency tables can have 25% of error for the third-order ones. Some variants are also described and compared in these papers.

A discussion on the contribution of Dalenius and Reiss can be found in [14].

6.1.1.1   Rank Swapping

The first proposal of rank swapping seems to be for ordinal data in [15]. Then, it was applied to numerical data in [16]. This last paper also proposes an enhanced version with the goal that the masked data set preserves (i) the multivariate dependence or independence in the original file, and (ii) the means of subsets. One of the reasons for introducing rank swapping is that arbitrary swapping "can severely distort the statistics of sub-domains of the universe" [16].

The definition of rank swapping is given in Algorithm 20. It is defined for a single variable. When the data file contains several variables, the method is applied to each of them independently. The algorithm depends on a parameter p that permits the

---

Algorithm 20: Rank swapping: $rs(X, p)$.

Data: X: original data file; p: percentage of records for swapping
Result: X': masked file
1 $(a_1, \ldots, a_n)$:= values of variable V in X in increasing order (i.e., $a_i \leq a_\ell$ for all
   $1 \leq i < \ell \leq n$) ;
2 Mark a_i as unswapped for all i ;
3 **for** $i = 1$ to n **do**
4 $\quad$ **if** a_i *is unswapped* **then**
5 $\quad\quad$ Select ℓ randomly and uniformly chosen from the limited range
   $\quad\quad$ $[i + 1, \min(n, i + p * |X|/100)]$;
6 $\quad\quad$ Swap a_i with a_ℓ ;
7 Undo the sorting step ;

---

user to control the amount of disclosure risk. Normally, p corresponds to a percent of the total number of records in X.

With this algorithm, the smaller the p, the larger the risk. Note that when p increases the difference between x_i and x_ℓ may increase accordingly. Therefore, the risk decreases. Nevertheless, in this case the differences between the original and the masked data set are higher, so information loss increases.

One of the main advantages of rank swapping is that the marginal distributions of the original file and the masked file are equal because the values are not modified. Also, the method gives a good trade-off between information loss and disclosure risk when the transparency principle is not taken into account. It was classified in [17] among the best microdata protection methods for numerical attributes and in [18] among the best for categorical attributes. We discuss in the next section the risk under the transparency principle.

One of the disadvantages of these methods is that correlations between the variables are modified by the masking method. This is due to the fact that variables are masked independently. This is one of the issues addressed by Moore in [16].

A discussion on earlier contributions to data and rank swapping can be found in [14]. References [19,20] are two of the existing variants of rank swapping. Reference [21] describes data shuffling a related approach.

6.1.1.2 Rank Swapping and Transparency: A Specific Attack

Recall that when we apply the transparency principle, we do not only publish the masked file but also give the information that the data has been masked with rank swapping and release the parameter p used. In this case, any attacker can use this information to attack the database.

In this section we describe a specific record linkage algorithm (rank swapping record linkage, RS-RL) that uses this particular information to link the records of the intruder with the ones in the released masked file. This method is described in detail in [22]. The procedure takes advantage that for any given value, swapping is constrained to a subset of records. Using the notation in Algorithm 20, the value a_i can only be swapped with values ℓ in the limited range $[i + 1, \min(n, i + p * |X|/100)]$.

Therefore, when an intruder knows the value of a variable V_j about an individual x_i a set $B_j(x_i)$ that contains all masked records which may be the masked version of x_i can be computed. Intruders can apply the same process to all variables they know, and then the original record should correspond to one in the intersection of all the sets $B_j(x_i)$ for all known variables j. That is, let $1 \leq j \leq c$ denote the indices of the variables known by an intruder. Then,

$$x'_\ell \in \cap_{1 \leq j \leq c} B_j(x_i). \tag{6.1}$$

This type of attack is an intersection attack. And $B(x_\ell) = \cap_{1 \leq j \leq c} B_j(x_i)$ is the anonymity set of x_ℓ.

When this set consists of a single record, this is a true match. Note that in this case, we know for sure that the link is correct because there is no uncertainty in the process

Algorithm 21: Rank swapping record linkage.

Data: $Y \subseteq X$: data file of the intruder; X': masked file; p: percentage of records for swapping
Result: linkage between Y and X'

1 $LP = \emptyset$;
2 **for** *each* $x_i \in Y$ **do**
3 $\quad\quad B(x_i) = \cap_{1 \le j \le c} B_j(x_i)$;
4 $\quad\quad x' = \arg\min_{x' \in B(x_i)} d(x', x_i)$;
5 $\quad\quad LP = LP \cup (x', x_i)$;
6 **return** (LP) ;

and we are absolutely sure that the record is in the intersection of the $B_j(x_i)$. I.e., it is a match. This match will be included in the reidentification measure $K.Reid(B, A)$ (see Eq. 5.2). When the intersection includes several records, we have uncertainty but only within the intersection set. In this case, we can apply e.g. distance-based record linkage to select one of the records of the intersection set. If the method is known but the parameter is not, [22] proposes to set an upper bound of the parameter. In case that the set is empty for any of the records, it means that the estimation of the parameter is too low and it should be enlarged. Note that for the correct parameter p, the intersection will never be empty.

Algorithm 21 describes the transparency attack for rank swapping. We call it rank swapping record linkage. The following example illustrates this algorithm.

Example 6.1 [22] Let us consider the original and masked files of Table 6.1. Variables $V_1, \ldots, V_4$ represent quasi-identifiers and variable V_5 represents a confidential variable that is not a quasi-identifier. Then, only variables $V_1, \ldots, V_4$ are masked using rank swapping. For convenience the original and masked files are aligned. I.e.,

Table 6.1 Example for rank swapping record linkage: original and masked files from [22]

Original file X					Masked file X'				
V_1	V_2	V_3	V_4	V_5	V_1'	V_2'	V_3'	V_4'	V_5
8	9	1	3	il_1	10	10	3	5	il_1
6	7	10	2	il_2	5	5	8	1	il_2
10	3	4	1	il_3	8	4	2	2	il_3
7	1	2	6	il_4	9	2	4	4	il_4
9	4	6	4	il_5	7	3	5	6	il_5
2	2	8	8	il_6	4	1	10	10	il_6
1	10	3	9	il_7	3	9	1	7	il_7
4	8	7	10	il_8	2	6	9	8	il_8
5	5	5	5	il_9	6	7	6	3	il_9
3	6	9	7	il_{10}	1	8	7	9	il_{10}

the ith record in the original file corresponds to the ith record in the masked file. In the masking process we used $p = 2/10$. Thus, each value could have been swapped with 4 other values (2 smaller and 2 larger values).

Let us consider that the intruder has information corresponding to variables V_1, V_2, V_3, V_4 for an individual and that this information is $(6, 7, 10, 2)$.

First, let us consider the standard distance-based record linkage method (with the Euclidean distance) applied to this record $(6, 7, 10, 2)$. We compute the distance between this record and all the masked records. The closest masked record is $(6, 7, 6, 3)$, which is not the matching one. Therefore, distance-based record linkage leads to an incorrect linkage $(6, 7, 10, 2) \leftrightarrow (6, 7, 6, 3)$.

Let us now consider the record linkage described in Algorithm 21 for the same record $(6, 7, 10, 2)$. The set of possible masked records consistent with the 6 in the first variable V_1 corresponds to $B_1(V_1 = 6) = \{(4, 1, 10, 10), (5, 5, 8, 1), (6, 7, 6, 3), (7, 3, 5, 6), (8, 4, 2, 2)\}$. Let us compute also the sets for the other variables. We obtain $B_2(V_2 = 7) = \{(5, 5, 8, 1), (2, 6, 9, 8), (6, 7, 6, 3), (1, 8, 7, 9), (3, 9, 1, 7)\}$, $B_3(V_3 = 10) = \{(5, 5, 8, 1), (2, 6, 9, 8), (4, 1, 10, 10)\}$ and $B_4(V_4 = 2) = \{(5, 5, 8, 1), (8, 4, 2, 2), (6, 7, 6, 3), (9, 2, 4, 4)\}$. Then, the intersection of these sets (Eq. 6.1) is a single record $(5, 5, 8, 1)$, which is the correct linkage.

Naturally, the more the variables known by the intruder, the easier that the intersection is a single record. The experiments reported in [22] show that this attack is quite effective in reidentifying the masked data set. The main characteristics of this approach are summarized below.

- Given a record, when the intersection set is a single record, this is a match. That is, we know for sure that this is the record we are looking for. Recall our discussion in Sect. 5.3.4 (Eq. 5.2 for $K.Reid$).
- Given a masked file, this attack can be done even in the case that the intruder only knows data from a single individual in the masked file. There is no need to have a set of data from several individuals to apply the attack.
- The more variables the intruder has, the better is the reidentification. The intersection set never increases when the number of variables increases.
- When the intruder does not know the parameter p, a lower bound of this parameter can be found. Given records in $Y \subseteq X$ known by the intruder, the lower bound of p corresponds to the minimum value that makes all anonymity sets B not empty. To formalize this lower bound of p (denoted p_{lw}), we denote the anonymity set of any record y computed using a particular p_0 by $B^{p_0}(y)$. That is, $B^{p_0}(y)$ is the set of records compatible with y when we are using rank swapping with parameter p_0. Then,

$$p_{lw} = \min\{p_0 | \ |B^{p_0}(y)| \geq 1, \text{ for all } y \in Y\}$$

Note that the larger the set Y, the larger the lower bound and, thus, the better the estimation of p.

6.1.1.3 Rank Swapping p-distribution and p-buckets

In order to overcome the problems of rank swapping record linkage, two variations of rank swapping were defined in [22]. They are p-distribution and p-buckets rank swapping. Both variations have in common that the set $B_j(x_i)$, using the terminology of Algorithm 20, is the whole data set. That is, swapping can be done with a value belonging to any record of the file. Probabilities are larger for records with similar values, and lower for records that are dissimilar but they are never zero. The difference between p-distribution and p-bucket rank swapping is on how the probabilities are defined.

In p-distribution rank swapping, the value of the ith record is swapped with the value of record $\ell = i + r$ where r is a random value computed using the $N(0.5p, 0.5p)$ normal distribution. Algorithm 22 describes the process. Here p is the parameter of the method and plays the same role as p in standard rank swapping.

In p-bucket rank swapping, ordered values are clustered into p buckets of similar size and with consecutive values. Let $B_1, \ldots, B_p$ denote these buckets. Then, for each value a_i of a bucket B_r we select a value a_l of a bucket B_s in a two step process. First, we select the bucket B_s. Selection is done according to the probability distribution

$$Pr[B_s \text{ is choosen } |B_r] = \frac{1}{K} \frac{1}{2^{s-r+1}}.$$

In this expression K is a constant so that the probabilities add up to one.

Once B_s is known, we select a value a_l from the bucket B_s. In this second step, the selection of a_l is done using a uniform distribution on the elements of the bucket. When $B_s = B_r$, $\ell > i$ is imposed.

p-Distribution rank swapping has been attacked recently in [23] using rank-based record linkage. The results show that p-Distribution rank swapping is still more resistant to transparency attacks than standard rank swapping.

Algorithm 22: Rank swapping p-distribution: $rs(X, p)$.

Data: X: original data file; p: percentage of records for swapping
Result: X': masked file

1 $(a_1, \ldots, a_n)$:= values of variable V in X in increasing order (i.e., $a_i \le a_\ell$ for all $1 \le i < \ell \le n$);
2 Mark a_i as unswapped for all i;
3 **for** $i = 1$ **to** n **do**
4   **if** a_i *is unswapped* **then**
5   Select r according to the normal distribution $N(0.5p, 0.5p)$;
6     $\ell = i + round(r)$;
7   Swap a_i with a_ℓ;

8 Undo the sorting step;

Alternatively, we can consider a distance based record linkage based on the OWA operator. Note that these two rank swapping variants ensure privacy granting that with a high probability at least one of the variables will be swapped with a value in a distant position. The distance based on the OWA permits to discard e.g. the largest distance. The combination of OWA, rank-based record linkage and supervised machine learning has not yet been considered for these rank swapping based methods. These combinations would improve the performance of rank-based record linkage.

6.1.1.4 Rank Swapping: Variants and Big Data

Rank swapping has been applied to ordinal categorical data. As order is the only property that matters, they have also been applied to numerical data. A variant of rank swapping on data with a partial order has also been considered [24]. This definition permits to apply rank swapping to several variables at once.

References [25, 26] introduce swapping for high-dimensional data. The approach is based on a spectral transformation of the data using singular value decomposition. I.e., data X is first expressed as UDV^T (where V^T is the transpose of V) and then a new file X' is computed considering a permutation $\tilde{U}$ of U and defining $X' = \tilde{U}DV^T$.

Swapping has been applied to location privacy in [27] in a distributed way. There are also methods for social networks based on swapping edges. As these methods can be seen as noise addition on the adjacency matrix of the graph, we include some references in the corresponding section of this chapter.

Reference [28] presents an approach for streaming data in the context of big data. The algorithm is based on a sliding window.

6.1.2 Microaggregation

Microaggregation [29] consists of building small microclusters and then replacing the original data in each cluster by the cluster representatives. Privacy is achieved because each cluster is required to contain a predefined number of records and, thus, the published data, the cluster representative, is not the data of a single record but a representation of the whole cluster.

This method is currently used by many statistical agencies [7]. Comparisons of microaggregation with other methods show that it gives a good trade-off between information loss and disclosure risk. See [17] for a comparison with other methods for numerical data and [18] for a comparison with other methods for categorical data.

Microaggregation is formulated mathematically as an optimization problem. The goal is to find the clusters, i.e. a partition of the data, that minimize a global error. The formalization of the problem follows the one of the k-means (see Sect. 2.3.1). Each cluster is represented with a characteristic function χ_i where $\chi_i(x) = 1$ if the record x is assigned to the ith cluster and $\chi_i(x) = 0$ if not. The clusters need to have a minimum number of records to satisfy the privacy requirements. This number, denoted by k in the optimization problem, is a parameter of microaggregation. The function to be minimized requires that the partition is made so that the distance

---

Algorithm 23: Optimal Univariate Microaggregation.

Data: X: original data set, k: integer
Result: X': protected data set

1 **begin**
2   Let $X = (a_1 \ldots a_n)$ be a vector of size n containing all the values for the attribute being protected. Sort the values of X in ascending order so that if $i < j$ then $a_i \leq a_j$.
3   Given A and k, a graph $G_{k,n}$ is defined as follows.
4   **begin**
5   Define the nodes of G as the elements a_i in A plus one additional node g_0 (this node is later needed to apply the Dijkstra algorithm).
6   For each node g_i, add to the graph the directed edges (g_i, g_j) for all j such that $i + k \leq j < i + 2k$. The edge (g_i, g_j) means that the values $(a_{i+1}, \ldots, a_j)$ might define one of the possible clusters.
7   The cost of the edge (g_i, g_j) is defined as the within-group sum of squared error for such cluster. That is, $SSE = \Sigma_{l=i+1}^{j} (a_l - \bar{a})^2$, where $\bar{a}$ is the average record of the cluster.
8   The optimal univariate microaggregation is defined by the shortest path algorithm between the nodes g_0 and g_n. This shortest path can be computed using the Dijkstra algorithm.
9   **end**
10 **end**

---

between records and their cluster centers is minimal. This is formalized by means of cluster centers p_i and a distance d between records and cluster centers.

The optimization problem includes a parameter c which is the number of clusters and two additional sets of constraints. One that requires that all records are assigned to one cluster, and another that states that the characteristic functions are either zero or one.

The definition of the problem is as follows.

$$\text{Minimize } SSE = \sum_{i=1}^{c} \sum_{x \in X} \chi_i(x)(d(x, p_i))^2 \qquad (6.2)$$

$$\text{Subject to } \sum_{i=1}^{c} \chi_i(x) = 1 \text{ for all } x \in X$$

$$2k \geq \sum_{x \in X} \chi_i(x) \geq k \text{ for all } i = 1, \ldots, c$$

$$\chi_i(x) \in \{0, 1\}$$

For numerical data it is usual to require that $d(x, p)$ is the Euclidean distance. For variables $\mathbf{V} = (V_1, \ldots, V_s)$, x and p are vectors, and $(d(x, p))^2 = \sum_{V_i \in V} (x_{v_i} - p_{V_i})^2$. In addition, it is also common to require for numerical data that p_i is defined as the arithmetic mean of the records in the cluster. I.e., $p_i = \sum_{j=1}^{n} \chi_i(x_j) x_j / \sum_{j=1}^{n} \chi_i(x_j)$.

In the case of univariate microaggregation, i.e. a single variable, using the Euclidean distance and the arithmetic mean, algorithms that find an optimal solution in polynomial time can be implemented. Algorithm 23 is one of such methods, it was introduced in [30]. In contrast, in the case of multivariate data sets the problem becomes NP-Hard [31]. For this reason, heuristic methods have been proposed in the literature.

6.1.2.1 Heuristic Methods

Heuristic methods are usually based on an operational definition of microaggregation. They are defined in terms of the following three steps.

- **Partition**. Records are partitioned into several clusters, each of them consisting of at least k records.
- **Aggregation**. For each of the clusters a representative (the centroid) is computed.
- **Replacement**. Each record is replaced by the cluster representative.

This operational definition has been used to extend microaggregation to a large number of data types. Note that given data in a certain domain, if we define a distance and an aggregation function we can then define a partition method to complete the three steps of the operational definition.

Both the general formulation and the heuristic operational definition permits us to apply microaggregation to multidimensional data. Nevertheless, when the number of attributes is large, it is usual to apply microaggregation to subsets of attributes; otherwise, the information loss is very high [32]. Algorithm 24 defines this procedure. It considers a partition $\Pi = \{\pi_1, \ldots, \pi_p\}$ of the variables in the data set X, and applies microaggregation to each subset $\pi \in \Pi$.

6.1.2.2 Strategies for Selecting Subsets of Attributes

Individual ranking is when we apply microaggregation to each attribute in an independent way. That is, applying Algorithm 24 with $\pi_i = \{V_i\}$.

Using a partition Π of the attributes permits us to reduce the information loss at the cost of increasing the disclosure risk. Different partitions can lead to differ-

---

Algorithm 24: General Multivariate Microaggregation.

Data: X: original data set, k: integer
Result: X': protected data set
1 **begin**
2 | Let $\Pi = \{\pi_1, \ldots, \pi_p\}$ be a partition of the set of variables $V = \{V_1, \ldots, V_s\}$
3 | **foreach** $\pi \in \Pi$ **do**
4 | ⌊ Microaggregate X considering only the variables in π
5 **end**

---

ent measurements of risk and information loss. It has been shown [33] that when correlated variables are grouped together, information loss is not as high as when uncorrelated variables are put together. However, the selection of uncorrelated attributes decreases disclosure risk and can lead to a better trade-off between disclosure risk and information loss. Reference [33] gives a detailed discussion with examples on the results for different partitions. Reference [34] considers the variable grouping problem as an optimization problem and uses genetic algorithms to solve it. The use of genetic algorithms for grouping problems as this one is presented in detail in [35].

A different approach is to microaggregate the given set of records using only a subset of available variables. This approach is considered in [36], where a set of variables is selected from the whole set according to the mutual information measure. In some sense, the selected set accounts for half the *independence* between the variables. This roughly corresponds to pick up variables that are not correlated, and use only these ones to build the clusters. Aggregation and replacement is applied to all the variables in the set. Note the difference between using the Pearson correlation and the mutual information measure. Reference [36] justify their approach on the dependency trees introduced by [37]. Reference [38] also study the problem of selecting variables and reach the same conclusion that information loss is decreased when independent variables are selected.

As a summary, if only information loss is under consideration, it is preferable to group together independent variables. When both disclosure risk and information loss are under consideration, what is preferable is not so clear.

6.1.2.3 Strategies for Building the Partition: Partitioning

For a given grouping of the set of variables $\Pi = \{\pi_1, \ldots, \pi_p\}$ we need to apply a multivariate microaggregation method to each subset of data with variables in π_i to partition, aggregate, and replace the value in the original data set. As stated above, a heuristic approach is usually applied.

Heuristic approaches for sets of attributes can be classified into two categories. One approach consists of projecting the records on a one dimensional space (using e.g. Principal Components or Zscores) and then applying optimal microaggregation on the projected dimension (Algorithm 26 describes this approach). The other is to develop adhoc algorithms. The MDAV [39] (Maximum Distance to Average Vector) algorithm follows this second approach. It is explained in detail in Algorithm 28, when applied to a data set X with n records and A attributes. The implementation of MDAV for categorical data is given in [40].

Reference [41] (only for bivariate data) and [42] introduced an heuristic method based on minimum spanning trees. The approach ensures that all clusters have at least k records but can contain more than $2k$ records unless an additional partitioning step is applied to this larger sets. The procedure is as follows:

- **Step 1**. F = minimum spanning tree(X)
- **Step 2**. F' = delete removable edges(F)
- **Step 3**. C = assign data to clusters(F')

The construction of the minimum spanning tree is done using the Prim's method. The nodes are the set of records in X, and the costs of the edges correspond to the (Euclidean) distance between records.

Once the minimum spanning tree of the whole data set X is built, some edges are removed in order to obtain smaller components (smaller trees) each with at least k elements. To do so, edges in the tree are considered from larger to lower length and if their removal is permitted (the two new smaller components have at least k records) the edge is cut.

Reference [42] gives a detailed description of this algorithm. It uses a priority queue of the edges based on their length, and also uses a node structure that links each node with its parent and that contains information on the number of descendents of the node. This latter information is useful when determining if an edge can be cut. Note that an edge linking a node o to a node d (i.e., o is the origin of the edge and d the destination as directed edges are used) can be removed when the number of descendents of o is at least k and the component of node d has also at least k nodes. That a component has at least k nodes can be checked traversing the nodes to the root and then comparing the number of descendents with k. This is detailed in Algorithm 25.

The minimum spanning tree heuristics can lead to clusters with more than $2k$ records when the removal of any edge of the cluster lead to less than k nodes. Such clusters have a star-shape, which are nodes with a central node connecting with several components all of them with less than k nodes. In order to reduce information loss (when measured using SSE) we should split such large clusters. Other microaggregation algorithms can be used for this purpose.

Reference [42] compares the performance of the algorithm with other algorithms and show that the best results are obtained when the clusters are well separated.

Reference [43] presents another heuristic microaggregation algorithm based on selecting clusters around points with high density. The density is computed taking into account the $k - 1$ nearest records of the record under consideration. This approach can be seen as a microaggregation version of the subtractive method [44]

Algorithm 25: Minimum spanning tree heuristic microaggregation: *isRemovable*$((o, d), F)$.

Data: (o, d): edge, F: minimum spanning tree
Result: Boolean
1 **begin**
2   **if** *descendents*$(o) < k$ **then**
3   return(false)
4   else root := obtain root (o, F)
5   return (descendents$(root)$ − descendents$(o) \geq k$)
6 **end**

Algorithm 26: Heuristic Multivariate Microaggregation: Projected Microaggregation.

Data: X: original data set, k: integer
Result: X': protected data set

1 **begin**
2   Split the data set X into r sub-data sets $\{X_i\}_{1 \le i \le r}$, each one with a_i attributes of the n records, such that $\sum_{i=1}^{r} a_i = A$
3   **foreach** ($X_i \in X$) **do**
4   Apply a projection algorithm to the attributes in X_i, which results in an univariate vector z_i with n components (one for each record)
5   Sort the components of z_i in increasing order
6   Apply to the sorted vector z_i the following variant of the univariate optimal microaggregation method: use the algorithm defining the cost of the edges $\langle z_{i,s}, z_{i,t} \rangle$, with $s < t$, as the within-group sum of square error for the a_i-dimensional cluster in X_i which contains the original attributes of the records whose projected values are in the set $\{z_{i,s}, z_{i,s+1}, \ldots, z_{i,t}\}$
7   For each cluster resulting from the previous step, compute the p_i-dimensional centroid and replace all the records in the cluster by the centroid
8 **end**

---

using a different potential function

$$d^k(x, T) = \frac{1}{\sum_{x \in N^k(x,T)} ||x - p||^2}$$

where $N^k(x, T)$ are the $k - 1$ nearest records to x in T and p is the average of these records. Here, T is the subset of X under consideration.

Heuristic methods lead to partitions that may be suboptimal with respect to the objective function SSE. Reference [45] proposes an additional step after clustering to improve the partition obtained by the clustering algorithm. It consists of reassigning the elements with a large variance. This is based on the fact that given a cluster C_i with a large contribution to the error (that is, $\sum_{x \in C_i} (d(x, p_i))^2$ is large), we may be able to reassign all the elements $x \in C_i$ to other clusters in such a way that

- (i) the clusters still have between k and $2k - 1$ records, and
- (ii) the total distance to the newly assigned cluster centers is smaller than the total distance to C_i. That is,

$$\sum_{x \in C_i} (d(x, p_{n(x)}))^2 < \sum_{x \in C_i} (d(x, p_i))^2$$

where $n(x)$ is the function that assigns the new cluster center to records in C_i.

Algorithm 27 details this process.

Algorithm 27: Reassignment of records to reduce *SSE*.

    Data: X: original data set, C: partition of X
    Result: C': a new partition

1 **begin**
2   Compute $GSE_i = \sum_{x \in C_i} (d(x, p_i))^2$ for each cluster $C_i \in C$
3   Define $GSE_{\sigma(i)}$ so that $GSE_{\sigma(i)} \geq GSE_{\sigma(i+1)}$
4     $C' = C$
5     $SSE_{C'} = SSE_C$
6   **for** $i = 1$ *to* $|C|$ **do**
7   **for** $x \in C_i$ **do**
8   Assign temporarily x in C' to $\arg\min_{i \neq \sigma(i)} d(x, p_i)$
9   Update C' and $SSE_{C'}$ accordingly
10   **if** $SSE_C > SSE_{C'}$ **then**
11           $C = C'$

12 **end**

6.1.2.4 Strategies for Aggregation

Given a partition, in principle, any aggregation function [46] can be used to build a representative of a cluster. For numerical data, the arithmetic mean is the most used aggregation function. The median and the mode (the plurality rule) have also been used. The median is a good option for ordinal data and the mode for nominal data.

6.1.2.5 Strategies for Replacement

Almost all approaches replace records by their cluster center. This corresponds to a deterministic replacement. Fuzzy microaggregation (discussed below) is one of the exceptions. It uses a random replacement approach that takes into account several cluster centers. This approach is used to avoid transparency attacks.

6.1.2.6 Microaggregation and Anonymity

Microaggregation can be used to obtain a masked file that is compliant with k-anonymity. This is achieved when the partition of the set of variables contains a set with all variables $\Pi = \{\pi = \{V_1, \ldots, V_s\}\}$. As other partitions of the variables do not ensure k-anonymity, [33] defined the *real anonymity measure* as the ratio between the total number of records in the original file and the number of protected records which are different:

$$k' = \frac{|X|}{|\{x'|x' \in X'\}|}. \tag{6.3}$$

Section 6.4 discusses the relation between both microaggregation and k-anonymity.

6.1.2.7 Postmasking for Microaggregation

There have been several attempts to improve the quality of microaggregated files decreasing the information loss of the protected data set. The approach in [47] (used in Example 5.6) replace different records of a cluster by different values so that the variance of the original and protected files are the same. Let C_i be the ith cluster, $|C_i|$ the cardinality of this set, and c_i the cluster center. Then, for each cluster C_i we define a sphere B_i with center c_i and radius r_i. Instead of replacing the records by the cluster center, we select $|C_i|$ points from the surface of B_i such that their average is the cluster center c_i. In order that the variance is maintained we select $r_i = \sqrt{Var(P_i)}$. Under these conditions, both mean and variance of the original set is preserved.

Blow-up microaggregation [48] is another way to keep both mean and variance of the original file. This is achieved moving all cluster representatives far away from the center of the data in such a way that the original and the protected data have the same variance.

Proposition 6.1 *Let X be the original database and X' be the protected database using standard microaggregation. Then, the database defined by*

$$\tilde{X} = X' + \beta \cdot (X' - \bar{X})$$

with

$$\beta = \sqrt{\frac{\sum (x_i - \bar{X})^2}{\sum_{k \in K} \alpha_k (\bar{x'}_k - \bar{X})^2} - 1}$$

has the same mean and variance as X.

Reference [49] proposes another approach (hybrid microdata using microaggregation) that combines microaggregation with synthetic data generators to preserve the covariances of the original data set.

Algorithm 27 discussed above to modify the partition to improve SSE can be seen as related to postmasking. Note however, that in this section we focused on the modification of the whole masked data and not only on the partition.

6.1.2.8 Microaggregation and Transparency

When data is protected using microaggregation visual inspection is enough to know that the file has been microaggregated, which is the parameter k and which is the partition of the variables used. This permits any intruder to build specific attacks to data masked using microaggregation.

Very effective attacks can be done for optimal univariate microaggregation. This is due to the fact that in optimal microaggregation, clusters are defined with contiguous values. Recall from Algorithm 23 that data is first ordered, and only sets with values $(a_{i+1}, \dots, a_j)$ are considered. Therefore, if $p_1, \dots, p_c$ are the cluster centers and z

the value in the hands of the intruder, if there is $p_i < z < p_{i+1}$, we know for sure that the record associated to z can only be either in the ith cluster or in the $(i + 1)$th cluster. Then, we define $B(z)$ as the set of records in the ith cluster union the set of records in the $(i + 1)$th cluster. In this way, we can attack the masked data set applying an intersection attack similar to the one we described for rank swapping (see Eq. 6.1). In this case using $B_j(x_i)$ to denote the set of masked records that correspond to a record x_i when we know the value of its V_j variable, we have that its masked version x'_ℓ should satisfy

$$x'_\ell \in \cap_{1 \le j \le c} B_j(x_i).$$

As in the case of rank swapping, this attack can be applied by an intruder even in the case of having information of a single record. Some non-optimal univariate algorithms also result into clusters with contiguous values. In this case, the same attack can be applied.

The following example illustrates a transparency attack.

Example 6.2 Let us consider the following one dimensional set of eight records $X = \{x_0 = 0, x_1 = 1.0, x_2 = 2.1, x_3 = 3.3, x_4 = 4.6, x_5 = 4.7, x_6 = 4.8, x_7 = 4.9\}$. If we apply a microaggregation with $k = 4$ we would obtain the clusters $\{0, 1.0, 2.1, 3.3\}$ and $\{4.6, 4.7, 4.8, 4.9\}$ with cluster centers $p_1 = 1.6$ and $p_2 = 4.75$. It is easy to see that if we look for the record $x_3 = 3.3$ the nearest cluster center is p_2 although it has been assigned to p_1.

- $d(x_3, p_1) = d(3.3, 1.6) = 1.7$
- $d(x_3, p_2) = d(3.3, 4.75) = 1.45$

When an intruder only considers distance-based record linkage for attacking the masked data set, x_3 is assigned to the wrong cluster. The transparency principle would permit the intruder to know that x_3 can be either assigned to p_1 or p_2. Further information on the data (e.g., other records and/or other variables) permits the intruder to further reduce the set of alternatives.

The partition obtained in this example would also be obtained by some heuristic microaggregation methods when $k = 3$. This is the case, for example, of the method based on minimum spanning trees. MDAV does not return this partition for $k = 3$.

Attacks for multivariate microaggregation are not so easy. Reference [50] presents a detailed analysis of such attacks. Different multivariate algorithms were attacked using a projected record linkage. For some projected microaggregation the improvement was very much significant (e.g., multiplying by three or even more the number of reidentifications of distance based and probabilistic record linkage). For MDAV algorithm, the increment was moderate but also significant. E.g, with a partition of variables into sets of 4 variables and with $k = 25$, the number of reidentifications increased from 11.20% for the generic distance based record linkage to 34.63% for the projected record linkage for one data file, and from 8.11 to 16.81% in the other data file considered.

In the literature we find a few other works in this topic. Reference [51] is the first to describe specific distances that are effective for record linkage for univariate microaggregation. Reference [52] was the first to discuss the problem for the multivariate case. In this work a microaggregation approach based on fuzzy clustering was introduced to avoid that intruders have and use this information. Reference [53] describes the intersection attack and experimentally evaluates its results for univariate microaggregation, and [50], as mentioned above, studies the case for a few algorithms of multivariate microaggregation.

6.1.2.9 Fuzzy Microaggregation

Fuzzy microaggregation has been proposed as a way to overcome transparency attacks. It is based on fuzzy clustering. In the partition step we have that records belong to different clusters with different membership degrees. This is used later in the replacement step as records can be assigned to different clusters according to a probability distribution based on the membership functions.

The first fuzzy microaggregation algorithm was introduced in [52,54,55]. Records are clustered by a fuzzy clustering algorithm that requires clusters to have at least k-records: Adaptive *at least k* fuzzy c-means. Then, replacement by the cluster center is done taking into account membership degrees.

Algorithm 28: MDAV algorithm.

Data: X: original data set, k: integer
Result: X': protected data set
1 **begin**
2     $C = \emptyset$
3   **while** $|X| \geq 3k$ **do**
4        $\bar{x}$ = the average record of all records in X
5   x_r = the most distant record from $\bar{x}$
6   x_s = the most distant record from x_r
7   C_r = cluster around x_r (with x_r and the $k-1$ closest records to x_r)
8   C_s = cluster around x_s (with x_s and the $k-1$ closest records to x_s)
9   Remove records in C_r and C_s from data set X
10       $C = C \cup \{C_r, C_s\}$
11   **if** $|X| \geq 2k$ **then**
12        $\bar{x}$ = the average record of all records in X
13   x_r = the most distant record from $\bar{x}$
14   C_r = cluster around x_r (with x_r and the $k-1$ closest records to x_r)
15        $C_s = X \setminus C_r$ (form another cluster with the rest of records)
16        $C = C \cup \{C_r, C_s\}$
17   **else**
18        $C = C \cup \{X\}$
19 **end**

Algorithm 29: Decoupled fuzzy c-means based microaggregation with parameters c, m_1, and m_2.

Data: X: original data set, k: integer, m_1, m_2: degrees of fuzziness
Result: X': protected data set

1 **begin**
2 | Apply fuzzy c-means with a given c and a given $m := m_1$
3 | For each x_j in X, compute memberships u to all clusters in $i = 1, \ldots, c$ for a given m_2. Use:

$$u_{ij} = \Big(\sum_{r=1}^{c} \Big(\frac{||x_j - v_i||^2}{||x_j - v_r||^2} \Big)^{\frac{1}{m_2 - 1}} \Big)^{-1}$$

4 | For each x_j determine a random value χ in $[0, 1]$ using a uniform distribution in $[0,1]$, and assign x_j to the ith cluster (with cluster center c_i) using the probability distribution $u_{1j}, \ldots, u_{cj}$. Formally, given χ select the ith cluster satisfying $\sum_{k<i} u_{kj} < \chi \leq \sum_{k \leq i} u_{kj}$. Define x'_j as the c_i

5 **end**

Recently, another simpler algorithm based on fuzzy c-means was introduced in [56,57]. The algorithm first clusters records using fuzzy c-means. In this case, no constraint is used on the number of records in each cluster, and only the usual parameters of fuzzy c-means are used: m_1 (the degree of fuzziness) and c the number of clusters. c is selected in the range

$$c \in \left[\frac{|X|}{2k}, \frac{|X|}{k} \right]$$

so that clusters have in average between k and $2k$ records.

Once the fuzzy partition is built, we compute the membership of each record to each cluster. For this purpose, we need to select a degree of fuzziness m_2 (which is not necessarily the same m_1 used for clustering). Then, records are assigned to clusters using a probability distribution built from the membership degrees. Algorithm 29 (from [57]) details these steps. We call this algorithm Decoupled fuzzy c-means based microaggregation. This is so because we apply fuzzy c-means using two different fuzziness degrees instead of one as usual.

With appropriate selection of c, m_1, and m_2 we obtain masked files that range from the original file to files with large information loss. In particular, the following can be proven. See [57] for details.

Proposition 6.2 *Decoupled fuzzy c-means based microaggregation satisfies the following properties.*

- *For large values of m_2 and $c = |X|/k$, the expected size of all clusters is k (probabilistically expected k-anonymity).*
- *For large m_2, all memberships tend to be $1/c$ and thus any replacement is equally probable.*

- *For large m_1, all cluster centers tend to collide $c_i = c_j = \bar{X}$.*
- *For $c = 1$, all records are replaced by $\bar{X}$.*

Here, recall that m_1, m_2 are values larger than 1. With a value near one, fuzzy c-means is equivalent to crisp k-means. In applications, it is usual to consider values between 1 and 2. With respect to this proposition, values larger than 2 are already considered large.

The first property means that we can have probabilistically expected k-anonymity. As we have that assignment is probabilistic, we cannot ensure that all clusters have the same number of records, but the expected number of records is the same for all clusters. The second result implies that we cannot always assume that records are replaced by the nearest cluster center. In fact, they can be replaced by any cluster center when m_2 increases. The last two results give conditions for the largest information loss possible (i.e., either $c = 1$ or a large m_1).

Reference [57] also discusses the problem of isolated records. They can cause large risk with small values of m_1 and m_2. In fact, with small values of m_1, we may have cluster centers that are just one (or a slightly modified version of one) record.

6.1.2.10 Microaggregation: Variants and Big Data

Some algorithms for microaggregation have been defined with the specific goal of being fast and thus usable for large volumes of data. See e.g. [45,58–61]. These methods focus on datasets with a large number of records. Alternatively, for methods for a large number of attributes see [25,26,62]. The former is based on the cosine distance and the later on spectral anonymization (i.e., to use a spectral basis in anonymization). In particular, the authors of the latter work transform data in another space using singular value decomposition.

Our discussion on microaggregation was focused on standard databases. Nevertheless, extensions have been defined for other types of databases. Note that the heuristic definition of microaggregation only requires an implementation of a partitioning algorithm (clustering), an aggregation method, and a way to do the replacement. As clustering algorithms have been defined for almost all types of data, we can use these algorithms to define microaggregation for any type of data.

For example, microaggregation algorithms exist for the following types of data.

- Time series. See e.g., [63]
- Location privacy. See e.g. [64–67].
- Graphs and social networks. Variations exist on whether we consider k-degree anonymity or a more strict privacy model. In the former case, we build a k-anonymous degree sequence [68,69] and then modify the original graph to be compliant with this sequence. In the latter case, one approach is to build clusters and then use these clusters as the new nodes—supernodes (including, when needed, k copies of each of them) and adding appropriate edges. See e.g. [70,71].

- Access and search logs. Microaggregation algorithms have been defined for both types of logs. See e.g. [72–75]. Some of these works use ontologies to take into account the meaning of the words in the queries.
- Documents represented as vectors. Some of the approaches consider only distances on vectors of frequencies while others used semantics for computing distances between terms.

6.1.3 Additive and Multiplicative Noise

As the name indicates, additive noise protects data adding noise into the original file. That is,

$$X' = X + \varepsilon,$$

where ε is noise following a certain distribution. The simplest approach is to require ε to be such that $E(\varepsilon) = 0$ and $Var(\varepsilon) = kVar(X)$ for a given constant k. The two main approaches are correlated and uncorrelated noise.

Uncorrelated noise addition corresponds to the case that for variables V_i and V_j, noise is such that besides of $E(\varepsilon) = 0$ and $Var(\varepsilon) = kVar(X)$ we have $Cov(\varepsilon_i, \varepsilon_j) = 0$ for $i \neq j$. In this case, additive noise preserves means and covariances, but neither variances nor correlation coefficients. Note that,

$$E(X') = E(X) + E(\varepsilon) = E(X)$$

$$Cov(X'_i, X'_j) = Cov(X_i, X_j) \text{ for } i \neq j$$

$$Var(X') = Var(X) + kVar(X) = (1 + k)Var(X)$$

$$\rho_{X'_i, X'_j} = \frac{Cov(X'_i, X'_j)}{\sqrt{Var(X'_i)Var(X'_j)}} = \frac{Cov(X_i, X_j)}{(1 + k)\sqrt{Var(X_i)Var(X_j)}} = \frac{1}{1 + k}\rho_{X_i, X_j}$$

In contrast, correlated noise addition preserves correlation coefficients and means. In this case, however, neither variance nor covariance is preserved: they are proportional to the variance and covariance of the original data set.

In correlated noise addition, ε follows a normal distribution $N(0, k\Sigma)$ where Σ is the covariance matrix of X.

$$E(X') = E(X) + E(\varepsilon) = E(X)$$

$$Cov(X'_i, X'_j) = (1 + k)Cov(X_i, X_j) \text{ for } i \neq j$$

$$Var(X') = Var(X) + kVar(X) = (1 + k)Var(X)$$

$$\rho_{X_i',X_j'} = \frac{Cov(X_i',X_j')}{\sqrt{Var(X_i')Var(X_j')}} = \frac{(1+k)Cov(X_i,X_j)}{(1+k)\sqrt{Var(X_i)Var(X_j)}} = \rho_{X_i,X_j}$$

These two types of additive noise illustrate a common problem in data privacy. It is difficult that the masked data satisfy all the properties we are interested in. When a new property is required to be satisfied in the masked data (e.g., correlation coefficients preserved) another property is lost (e.g., covariances). The method also illustrates that the transparency principle can help the user to analyse the data. In both types of noise addition, our knowledge on the method and the parameter k permits us to compute correct and exact values for the four statistics listed above.

First extensive testing of noise addition was due to Spruill [76]. Fuller [77] further studies this approach. Reference [78] gives an overview of these approaches for noise addition as well as more sophisticated techniques. Reference [79] also describes some of the existing methods as well as the difficulties for its application to privacy. These papers come from the SDC community.

Agrawal and Srikant [80] proposed later (in 2000) the use of noise addition in the PPDM community. They also discuss the reconstruction of the original distribution. Reference [81] also studies the reconstruction of the original data from the protected data. In their work two methods to reconstruct the data are proposed, one of them based on principal components analysis (PCA).

6.1.3.1 Multiplicative Noise

Multiplicative noise consists of defining X' as the product of the error by the original data. That is, $X' = X \cdot \varepsilon$. Different strategies can be used for the error. Reference [23] consider ε drawn from a uniform distribution on $[1 - b, 1 + b]$ where b is the parameter of the perturbation level. Within differential privacy it is customary to use Laplace distributions in the context of multiplicative noise. One of the differences between Laplace and normal distributions is that the former has heavier tails. See e.g., [82,83] for details on multiplicative noise.

An advantage of multiplicative noise with respect to additive noise is that the magnitude of the perturbation (the error) applied to a value is proportional to this value. That is, small values have in general small perturbation and large values have large perturbation.

6.1.3.2 Additive and Multiplicative Noise: Variants and Big Data

These methods can be easily implemented for numerical data. Definition for other types of data (categorical or structured data) is not so simple. In general, for noise addition, we need to define X' in terms of adding some noise ε from a given distribution on the domain associated to X.

For categorical data, the PRAM method (see Sect. 6.1.4) can be seen as the most natural alternative. Reference [84] provides another method when data consist of terms of text. It introduces semantic noise, that takes into account the semantics of the terms (semantics is based on an ontology).

Standard additive and multiplicative noise is well suited for large dimensions as the procedure is record-based. Therefore, the application of this masking method can be done on linear time with respect to the number of records. It can also be applied efficiently to streaming data. Note also that it is also well suited for records of high dimension.

Correlated noise addition needs the covariance matrix. This is problematic for a large dimension data files when correlation between variables change with respect to time. Otherwise, we can select an initial large subset of records, estimate the covariance matrix, and use it to protect the whole file. In this case, the approach is also cost-efficient once the correlation matrix has been built.

Different forms of additive noise have been applied to data sets of large dimensions. For example, to graphs and social networks. In this case, addition and removal of edges, as well as addition and removal of nodes have been considered. See e.g. [85,86].

6.1.4 PRAM

PRAM, Post-RAndomization Method [87], is a method for categorical data where categories are replaced according to a given probability. The method is usually applied to each variable independently.

Formally, PRAM is based on a Markov matrix on the set of categories. Let $C = \{c_1, \ldots, c_c\}$ be the set of categories, then P is a Markov matrix on C when $P : C \times C \rightarrow [0, 1]$ such that $\sum_{c_j \in C} P(c_i, c_j) = 1$. Then, X' is constructed from X replacing, with probability $P(c_i, c_j)$, each c_i in X by a c_j. Formally, the matrix of probabilities can be seen as a matrix of conditional probabilities. That is,

$$P(c_i, c_j) = P(X' = c_j | X = c_i).$$

The application of PRAM requires an adequate definition of the probabilities $P(c_i, c_j)$. Different matrices lead to different information loss and different disclosure risk. Naturally, maximum risk corresponds to the case of using the identity matrix because in this case $X' = X$. The literature presents different studies on how the Markov matrix can be selected. Reference [87] proposes invariant PRAM. Given $T_X = (T_X(c_1) \ldots T_X(c_c))$ the vector of frequencies of categories in C in file X, invariant PRAM consists of defining P in such a way that frequencies do not change. That is, $\sum_{i=1}^{c} T_X(c_i) p_{ik} = T_X(c_k)$ for all k. Then, assuming without loss of generality that c_k is the category such that $T_X(c_k) \geq T_X(c_i)$ for all i, and given a parameter θ such that $0 < \theta < 1$, $p_{ij} = P(c_i, c_j)$ is defined as follows:

$$p_{ij} = \begin{cases} 1 - \frac{\theta T_X(c_k)}{T_X(c_i)} & \text{if } i = j \\ \frac{\theta T_X(c_k)}{(c-1) T_X(c_i)} & \text{if } i \neq j \end{cases}$$

We can observe that for $i \neq j$ we have that

$$p_{ij} = \theta T_X(c_k)/((c-1)T_X(c_i)) = \frac{1 - p_{ii}}{c - 1}.$$

So, given a category, the probability of changing it is equally divided among all other categories. I.e., $p_{ij} = p_{ik}$ for all $j, k \neq i$.

Note that a θ equal to zero implies no perturbation, and θ equal to 1 implies total perturbation. So, θ permits the user to control the degree of distortion suffered by the data set.

Reference [88] presents an alternative approach for defining the Markov matrix. In this case, the probabilities p_{ij} for $i \neq j$ are defined by

$$p_{ij} = \frac{(1 - p_{ii})(n - T_X(i) - T_X(j))}{(n - 2)(n - T_X(i))} \tag{6.4}$$

where $T_X(i)$ is the frequency of category c_i and n is the number of records or the dimension of the file (i.e., $n = \sum_{k=1}^{c} T_X(k)$) and p_{ii} are predefined constant values for all i. It is not required all p_{ii} be the same.

It can be proven that $p_{ii} + \sum_{j=1 \, j \neq i}^{c} p_{ij} = 1$ for all i. The proof follows.

$$p_{ii} + \sum_{\substack{j=1 \\ j \neq i}}^{c} p_{ij} = p_{ii} + \sum_{\substack{j=1 \\ j \neq i}}^{c} \frac{(1 - p_{ii})(n - T_X(i) - T_X(j))}{(n - 2)(n - T_X(i))}$$

$$= p_{ii} + \frac{(1 - p_{ii})}{(n - 2)(n - T_X(i))} \sum_{\substack{j=1 \\ j \neq i}}^{c} (n - T_X(i) - T_X(j))$$

$$= p_{ii} + \frac{(1 - p_{ii})}{(n - 2)(n - T_X(i))}((c - 1)(n - T_X(i)) + \sum_{\substack{j=1 \\ j \neq i}}^{c} T_X(j))$$

$$= p_{ii} + \frac{(1 - p_{ii})}{(n - 2)(n - T_X(i))}((c - 1)(n - T_X(i)) + (n - T_X(i)))$$

$$= p_{ii} + \frac{(1 - p_{ii})}{(n - 2)(n - T_X(i))}(c - 2)(n - T_X(i))(n - T_X(i)) = 1$$

The key point of this equation is that it assigns the higher exchange probabilities to the categories with less frequency. This definition is to increase confusion and reduce the risk of reidentification for the records with these categories that have, in fact, a larger probability of being unique in the file.

Reference [89] proposes the computation of matrix P from a preference matrix $W = \{w_{ij}\}$ where w_{ij} is our degree of preference about replacing category c_i by category c_j. Formally, given W the probabilities P are determined from the following optimization problem.

$$\text{Minimize} \quad \sum_{i,j} w_{ij} p_{ij} \tag{6.5}$$

$$\text{Subject to} \quad p_{ij} \geq 0$$

$$\sum_{j} p_{ij} = 1$$

$$\sum_{i=1}^{c} T_X(c_i) p_{ij} = T_X(c_j) \text{ for all } j$$

Reference [89] use integers to express preferences, and $w_{ij} = 1$ is the most preferred change, $w_{ij} = 2$ is the second most preferred changes, and so on.

The problem of finding an optimal PRAM matrix for a given data set has been considered in the literature. Given a data file, the goal is to obtain a Markov matrix that leads to a file with minimum disclosure risk and minimum information loss. In [90] an approximation of the optimal solution was computed using genetic algorithms. In [91] the same approach is applied requiring the Markov matrix to satisfy invariance (i.e., optimal invariant PRAM matrix). In [92] the problem was addressed using genetic programming. In this case, instead of looking for a Markov matrix, the method looks for an analytical expression that permits to compute the matrices.

6.1.4.1   PRAM and Transparency

Transparency attacks can also be defined for PRAM. These attacks, as they did for microaggregation and rank swapping, will exploit the fact that variables are masked independently. Similarly to what happens for rank swapping and microaggregation, attacks will be effective when the possible values of a given original value is a small subset of the domain. Using the same notation we used previously (i.e., $B_j(x)$ denotes the anonymity set for the jth variable for x), we have that when $B_j(x) \subset X'$ we may have that $\cap B_j(x)$ can be reduced to a single record and reidentification takes place. When $P(c_i, c_j) \neq 0$ for all i, j we have that $B_j(x) = X'$ and the intersection attack is not so effective. Note, however, that PRAM permits to compute accurate probabilities of transforming a record $x \in X$ (with several variables) into a record $x' \in X'$.

6.1.4.2   PRAM: Variants and Big Data

PRAM is defined for nominal data but can be used for other types of data (e.g. continuous variables) provided that we can define properly the Markov matrix. Note that it is not necessary that the Markov matrix is explicitly defined. That is, given a value c_i in a domain D, and a random value $r \in [0, 1]$ it is enough to have a function f such that $f(c_i, r)$ returns c_j (the masked value of c_i in D). Typically, for a Markov matrix $P(c_i, c_j)$ this function is defined as follows: if $r \in [0, P(c_i, c_1)]$ we have c_1, if $r \in [P(c_i, c_1), P(c_i, c_2)]$ we have c_2 and so on.

In a more general setting, we can consider $P(c_i, c_j)$ as $P(c_j|c_i)$ which is a valid expression for a probability distribution for any domain D. That is, not solely for categorical data. This formulation links PRAM with other masking methods as additive and multiplicative noise, and shows a way to apply PRAM to types of data different to categorical data. For example, we can apply PRAM to linguistic terms (e.g., words) where the domain of terms D is an ontology or a dictionary as wordnet in which relationships between terms are explicitly given. Then, we define the probability of replacement as uniform between all words at a given distance. Then, we select a word at distance 1 with probability $1/2^2$, at a distance 2 with probability $1/2^3$, and at a distance i with probability $1/2^{i+1}$. Not making any replacement should be $1 - \sum_i 1/2^{i+1}$.

For big data, PRAM is similar to noise addition in the sense that it is well suited for files of large volumes. Once Markov matrices are settled for all variables, masking is done record-wise. Therefore the cost of the approach is linear with respect to the number of records. The method also applies well to streaming data.

6.1.5 Lossy Compression and Other Transform-Based Method

Lossy compression was first proposed as a method for data privacy in [3]. The approach consists of viewing a numerical data file as a grey-level image. Then, a lossy compression method is applied to the *image*, obtaining a *compressed image*. This *image* is then decompressed and the *decompressed image* corresponds to the masked file.

Different compression rates lead to files with different degrees of distortion. I.e., the more compression, the more distortion. Reference [3] used JPEG, which is based on DCT, for the compression. References [93,94] uses JPEG 2000, which is based on wavelets.

The transformation of the original data file into an *image* requires, in general, a quantization. As JPEG 2000 allows a higher dynamic range (65536 levels of grey because it uses 16 bits of depth) than JPEG (only 256 because it uses 8 bits), this quantization step is more accurate in JPEG 2000.

Different quantization approaches are compared in [93,94]. Linear quantization transforms linearly the range of the original variables into the image range [0, 255] or [0, 65536]. Non uniform quantization can assign more values in the image space to those regions in the original domain that are more dense. A non-uniform quantifier based on histogram equalization [95] was used in [94]. While, in general, methods based on lossy compression lead to bad results with respect to the trade-off between information loss and disclosure risk, [94] shows that some well tunned masked files can be obtained that are comparable to those obtained with more effective methods (as rank swapping and microaggregation).

Independently, Bapna and Gangopadhyay introduced in [96] the use of wavelets in similar terms. The approach is based on the decomposition of a matrix using wavelets. In this work, the authors apply a row reduction (reduction on the number of individuals). Several works follow a similar approach. For example, [97,98] use also wavelet decomposition. The authors apply a function to reduce high frequency "noise". Then the inverse discrete wavelet transformation is applied to transform the data to the original space. Reference [98] applies a normalization approach so that the mean and standard deviation is preserved in the protected file. References [99,100] are similar approaches using in this case Haar wavelet transforms and focusing on other types of data uses (clustering instead of e.g. classification).

Another approach has been the use of Fourier-related transforms. Mukherjee et al. introduced this approach in [101].

The approach of transforming the original dataset into another space in order to reduce the noise or to apply a dimensionality reduction, has been considered in other contexts. For example, [102,103] introduce an approach based on the singular value decomposition (SVD), and [104] on the nonnegative matrix factorization. The approach for microaggregation and rank swapping working on the spectral basis (described in [25,26] and cited in Sects. 6.1.1 and 6.1.2) can also be seen from this perspective.

6.1.5.1 Lossy compression and transparency

From the point of view of transparency, lossy compression seems difficult to attack unless all the original file is already known by the intruder and, thus, all masking process is reproduced. This difficulty is due to the fact that the transformation suffered by any record depends strongly on all the other records in the original file.

The problem of determining the original file is considered in [96,97]. The authors use the term breaching algorithm for this type of attack. The goal is to find the records that can generate the protected file. It is shown that finding the original records can be easy when data is binary. The papers discuss the effectiveness of the approach in the more general case of real data.

6.1.5.2 Lossy Compression: Variants and Big Data

We have discussed lossy compression and related methods. These methods have not been applied to other data than numerical and categorical (see e.g. [96]). Nevertheless, we can apply lossy compression to any data file that can be taken as an image.

Reference [105] discusses the use of wavelet transforms for differential privacy of range-count queries.

6.2 Non-perturbative Methods

In this section we review some of the non-perturbative algorithms. We review generalization (also known as recoding), top and bottom coding, and suppression. Sampling [4], which is not discussed here, can also be seen as a non-perturbative method. Recall that sampling consists of selecting some records from the whole population. See e.g. [106, 107].

6.2.1 Generalization and Recoding

This method is mainly applied to categorical attributes. Protection is achieved by means of combining a few categories into a more general one. Local and global recoding can be distinguished.

Global recoding [4, 108] corresponds to the case that the same recoding is applied to all the categories in the original data file. Formally, if Π is a partition of the categories in C, then each c in the original data file is replaced by the partition element in Π that contains c.

In contrast, in local recoding (see e.g. [20]) the same category might be replaced by different generalizations when found in different records. Constrained local recoding is when the data space is partitioned and within the region the same recoding is used, but different regions use different recodings.

In general, global recoding has a larger information loss (as changes are applied to all records without taking into account whether they need them to ensure privacy) than local recoding. Nevertheless, local recoding generates a data set that has a larger diversity on the terms used to describe the records. This situation might cause difficulties when using the protected data set in analysis.

While most of the literature considers the recoding as functions of a single variable (i.e., single-dimension recoding), it is also possible to consider recoding of several variables at once. This is multidimensional recoding [109]. Formally, when n variables are considered, recoding is understood as a function of n values. We review a method for multidimensional recoding (Mondrian [109]) in Sect. 6.4.1.

One of the main difficulties for applying recoding is the need for a hierarchy of the categories. In some applications such hierarchy is assumed to be known (e.g. there exists an ontology or can be easily inferred from the semantics of terms), while in others it is constructed by the protection procedure. In this later case, difficulties appear because we want to determine the optimal generalization structure. Selection of an optimal generalization can be seen as equivalent to the selection of an optimal partition or an optimal dendrogram, depending on whether we need one or more levels of generalization. The number of partitions of a given set of n categories is the Bell number (see Sect. 2.3.1) and the number of dendrograms is $n!/2$. Because of that, in general, to find an optimal generalization for data protection is a hard problem and besides of that, it is preferable that the constructed hierarchy has semantic interpretation. E.g. generalizing Zip codes 08192, 08235, 09398, and 09247 into 0**9* and 0*2** may be inappropriate as these *codes* may correspond to sets of non-adjacent towns. Instead, it would be preferable to generalize them in 08*** and 09***.

Generalization has been extensively used to achieve k-anonymity. We will discuss these algorithms in Sect. 6.4.

6.2.1.1 Top and Bottom Coding

Top and bottom coding are two methods that consist of replacing the lowest and largest values (given a threshold) by a generalized category. These two methods can be considered as particular cases of global recoding, and as such are classified here.

These methods are applied when there are only a few records that have extreme values. This kind of generalization permits us to reduce the disclosure risk of outliers as the corresponding records are all generalized to the same category and are indistinguishable.

6.2.2 Suppression

Suppression consists of replacing some values by a special label denoting that the value has been suppressed. Suppression is often applied [110, 111] in combination with recoding, and mainly for categorical data.

We can have local and global supression. Global suppression is when suppressing a value c in record r implies that all other appearances of c in the file are also suppressed. Local suppression is when the suppression of c in one record is independent of the action on the values c in other records.

Suppression is often combined with generalization to achieve k-anonymity. This is the case of the Datafly algorithm [112]. We discuss algorithms for achieving k-anonymity in Sect. 6.4.

6.3 Synthetic Data Generators

Perturbative and non perturbative approaches are masking methods that try to ensure confidentiality modifying the original data. Synthetic data generators try to ensure confidentiality replacing the original data by artificial ones generated from models of the original data.

There are different approaches to generate synthetic data. Figure 6.1 outlines the requirements for synthetic data generators. Reference [113] reviews some of these methods. They distinguish three major classes.

- **Synthetic reconstruction**. Methods use a data set with some marginal distribution for the whole population, and then another data set (a sample) with the variables of interest for some individuals. The approach estimates first a joint distribution combining the marginals and the information from the sample, and then selects individuals from the sample in a way that is consistent with the joint distribution. The iterative proportional fitting (IPF) procedure is an old method (developed at the late 1930s) for building such set. Synthetic reconstruction has been used in [116].

- **Combinatorial optimization**. Methods use a data set with the variables to be simulated, and contingency tables for the variables. The method selects first a random sample from the file of the appropriate size. An iterative process is applied replacing a record in the selection by another in the file and evaluating the suitability of the replacement using a goodness-of-fit statistics (a fitness function using machine learning jargon).
- **Model-based simulation**. The process is based on the construction of models and the use of these models to generate the data. This approach is the one most studied for data protection in both fully and partially synthetic data sets, and it is the one we discuss below. We also use this approach in Algorithm 30.

The seminal work of synthetic data generators for data privacy are the works by Little [117] and Rubin [118]. Little's approach corresponds to partially synthetic data sets as only sensitive values (target variables in [117]) or identifiers (key variables in [117]) are synthetic. The second approach corresponds to fully synthetic data sets as all the values in the release are synthetic. Fully synthetic data sets are defined in terms of independent samples of records. In contrast, partially synthetic ones comprises the original records. In both cases, approaches differ on the methods used to find the model.

Algorithm 30 outlines how to generate partially synthetic data. The approach considers two types of variables X and Y. The algorithm builds synthetic versions of Y using models of Y with respect to X. Once the model is built, we replace Y by synthetic data generated using the model.

In fully synthetic data we consider three sets of variables (X, Z, Y) following [118]. We consider a microdata file of n records drawn from a larger population of N individuals. X represents background variables observed for all N individuals, Z represents non confidential variables observed for the n samples, and Y represents the confidential variables observed also for the n samples. First, a model of (Z, Y) is built in terms of the n records in X that are in the sample. Then, the model is used to impute values for the population in X. In order to avoid reidentification

---

Conditions for simulated data.

- Reflect actual sizes of regions and strata.
- Keep marginal distributions and interactions between variables.
- Keep heterogeneities between subgroups, especially regional aspects.
- Avoid pure replication of units from the underlying sample.
- Ensure data confidentiality.

Fig. 6.1 Conditions required for simulated data from [113], based on [114,115]

Algorithm 30: Partially synthetic data.

  Data: $X|Y$: set of records of a given sample
  Result: $X|Y'$: set of records with Y' a masked version of Y
1 $M_{X,Y} :=$ Build a model of Y in terms of X ;
2 $Y' := M_{X,Y}(X)$;
3 **return** $(X|Y)$;

---

problems [118] suggest imputation from the $N - n$ records (i.e., the records that are not in the sample). The same process can be used several times to create M different copies of the synthetic data set.

The type of synthetic data generator has implications for disclosure risk, which is larger for partially synthetic data. The more accurate is the model, the more similar are the synthetic records to the original ones. Effective reidentification of partially synthetic data sets generated using the IPSO method is described in [119]. For fully synthetic data no reidentification experiment has been reported in the literature, nevertheless discussion of disclosure risk for fully synthetic data is discussed in [120] (p. 408):

> When the imputation models are highly detailed, the imputations could reproduce combinations of quasi-identifiers for real records. Intruders might interpret this to mean that real-data records with those characteristics were in the original sample, which could result in identification disclosures if some of those records are unique in the population. This risk could be magnified when releasing multiple synthetic data sets, because (i) there are several opportunities to impute such records, and (ii) there could be repetitions of realistic synthetic records that might strengthen the intruder's confidence that a similar record was in the original data.

In addition, [121] discusses predictive disclosure for synthetic data, that can be seen as a type of attribute disclosure.

The use of synthetic data generators permits us to release multiple protected data sets. This permits users to increase the accuracy of their analysis (e.g., increasing the number of replicates reduces the variability in estimators of variance). See e.g. [121].

Reference [122] is a reference book on this area. A comparison of both fully synthetic and partially synthetic approaches is given in [123]. Some of the approaches used for synthetic data generation are the Information Preserving Statistical Obfuscation (IPSO) [124], enhanced GADP [125], methods based on the Latin Hypercube Sampling (LHS)—see e.g. [126]. The Enhanced GADP method is described in Algorithm 31. The algorithm includes the computation of Y_A, Y_B and Y_C. The algorithm returns Y_C that is the model that satisfies more properties (the matrices of regression coefficients $\hat{B}$ and covariances $\hat{\Sigma}$ are preserved). Y_A does not preserve them, and Y_B preserves only the regression coefficients. We can see them as three types of synthetic data with more or less precision, similar to IPSO-A, IPSO-B, and IPSO-C as discussed in [119].

While most methods for synthetic data generation are based on *statistical* models, there are some based on nonparametric methods from the machine learning literature. A reference on this approach, including an empirical evaluation, can be found in [127].

Algorithm 31: Enhanced GADP.

Data: Y: confidential; S: Non-confidential

Result: $Y'|S$: set of records with Y' a masked version of Y

1 $\bar{Y} :=$ mean vector of Y ;

2 $\bar{S} :=$ mean vector of S ;

3 $\hat{\Sigma}_{YY}, \hat{\Sigma}_{SS}, \hat{\Sigma}_{YS} :=$ covariance matrices of Y, S and between Y and S ;

4 Regress Y on S as follows

5 **begin**

6       $\hat{\beta}_1 := \hat{\Sigma}_{YS} \hat{\Sigma}_{YS}^{-1}$;

7       $\hat{\beta}_0 := \bar{X} - \hat{\Sigma}_{YS} \hat{\Sigma}_{SS}^{-1} \bar{S}$;

8       $\hat{\Sigma}_{\varepsilon\varepsilon} :=$ covariance of the residuals ;

9 **end**

10 $Y_A := \beta_0 + \beta_1 S$;

11 $K :=$ a $(n \times M)$ matrix of random numbers from a standard multivariate normal distribution ;

12 Regress K on $(S$ and $X)$. $B :=$ the residuals from this regression. ;; The new noise term B is orthogonal to both X and S and $\hat{B} = 0)$;

13 $Y_B := Y_A + B$;

14 $\hat{\Sigma} :=$ covariance matrix of B ;

15 Define C and Y computing for $i = 1, 2, \ldots, n$

16 **begin**

17       $c_i := \hat{\Sigma}_{\varepsilon\varepsilon}^{0.5} \hat{\Sigma}_{BB}^{-0.5} b_i$;

18       $y_i := \hat{\beta}_0 \hat{\beta}_1 s_i + c_i$;

19 **end**

20 $Y_C := Y$;

21 **return** $(X|Y)$;

The authors compare classification and regression trees, bagging, random forests, and support vector machines. The authors conclude

> that synthesizers based on regression trees are a particularly attractive option for statistical agencies seeking to release datasets with intense synthesis without intense labor.

6.4 Masking Methods and k-anonymity

In the literature there exist different approaches for achieving k-anonymity. Most of them are based on generalization and suppression. It has been proven that optimal k-anonymity with generalization and suppression is an NP-Hard problem (see e.g., [128–130]). Because of that some heuristic algorithms have been defined.

Another approach for achieving k-anonymity is the use of microaggregation methods. Microaggregation has to be applied considering all the variables at once, otherwise, k-anonymity would not be guaranteed. We have already seen microaggregation algorithms in Sect. 6.1.2.

As once k is established k-anonymity presumes no disclosure, only information loss measures are of interest here. Given a value for k, heuristic algorithms for k-anonymity look forward a k-anonymous file that has a minimum information loss.

6.4.1 Mondrian

Mondrian was proposed in [109] as a way to achieve k-anonymity for records described in terms of several variables. The method uses local recoding, and is based on the construction of a multidimensional partition. This partition defines a set of potentially overlapping regions, which are used for the recoding.

A greedy partitioning algorithm is used for achieving k-anonymity. The algorithm starts with a set of records. Then, it selects a variable to partition, and a value in the domain of that variable to partition the set of records in two sets of the same size. Then, the same algorithm is applied recursively to each subset. Recursion is stopped when no further partition is needed (or possible). Algorithm 32 formalizes this process.

Partition of elements is based on the selection of a value in the domain of a variable. This value (i_0 in the algorithm) is selected so that half of the records in the set (X in the algorithm) are in one of the subsets, and half in the other. That is, i_0 is the median of the values in X for variable V_i. As the domain may have records with the same value, it is possible that there are several records with the value i_0. These records are distributed in the two subsets to make them of the same size (or sizes differing in only one record).

Reference [109], following the description in [131], uses the variable with the largest spread in values for selecting the variable to split. When a set is not partitionable because it contains between k and $2k$ records, or because all records are identical, we return a recoding function γ that assigns to all records in the set X a summary of the set X.

Algorithm 32: Mondrian for achieving k-anonymity: *Mondrian*(X, k).

Data: X: original data set; k: integer
Result: X': protected data set
1 **if** *not(partitionable(X))* **then**
2 $\quad$ **return** $\{\gamma(x) = \{x \rightarrow summary(\mathrm{X})\} | x \in X\}$;
3 **else**
4 $\quad$ V_i = select variable from X ;
5 $\quad$ i_0 = select a value from domain of V_i in X ;
6 $\quad$ $lhs = \{x \in X | V_i(x) < i_0\}$;
7 $\quad$ $rhs = \{x \in X | V_i(x) > i_0\}$;
8 $\quad$ Distribute records in $\{x \in X | V_i(x) = i_0\}$ between lhs and rhs ;
9 $\quad$ **return** Mondrian(lhs, k) $\cup$ Mondrian(rhs, k) ;

6.4.2 Algorithms for k-anonymity: Variants and Big Data

In order to define algorithms for achieving k-anonymity in an efficient way, some methods are based on clustering. They tend to ressemble (or are equivalent to) microaggregation approaches. An example of such clustering-based approach for k-anonymity is given in [132]. Another, specific to location privacy, is [65]. For microaggregation algorithms for large dimensions we refer the reader to Sect. 6.1.2.

Some methods have been defined with the goal of improving first algorithms to achieve k-anonymity (i.e., Samarati's method [133], Datafly [112], Incognito [108], and Mondrian [109]). For example [134] wants to improve the algorithm in [133,135] describes the parallelization of Mondrian [109] using the map-reduce paradigm. Results about efficient implementations of clustering and of k-nearest neighbor (as e.g. [136]) are useful for efficient implementations of k-anonymity.

The k-anonymity model has been applied to all types of data, including time series, location privacy, graphs and social networks, acess and search logs, and documents. We add here some references not based on microaggregation but mainly based on generalization. For example, [27,137–139] are for location privacy, and [140] for social networks.

6.4.2.1 The Case of Social Networks

In the case of graphs and social networks, there is discussion (or disagreement) on the type of information available to intruders. This discussion implies that different sets of quasi-identifiers are used, which implies that different type of data protections can be applied. We call k-degree anonymity when we consider that the information available to an intruder is the degree of a node. Reference [141] was one of the firsts to consider this problem. Reference [142] considered the 1-neighborhood anonymity. In this case, the intruder has information on the neighbors of a node and the relationship between these nodes (i.e., edges between them). If $\mathcal{N}(n)$ is the graph of neighbors of n and their relationships, we have 1-neighborhood anonymity if for each node n there are $k - 1$ other nodes $n_1, \ldots, n_{k-1}$ such that the graphs $\mathcal{N}(n_i)$ are isomorphic. Reference [143] defines k-neighbor anonymity focusing only on the set of neighbors.

Reference [143] also introduces (k, l)-anonymity. A graph is (k, l)-anonymous if it is k-neighbor anonymous with respect to any subset of cardinality at most l of the neighbor sets of the vertices of the graph. There are two ways to interpret this: one in terms of the neighbors and the other taking also advantage of the non-neighbors. We have the following definition, in case that the non-neighbors are not considered.

Definition 6.1 Let $G = (V, E)$ be a graph. We say that G is (k, l)-anonymous if for any vertex $v_1 \in V$ and for all subset $S \subseteq N(v_1)$ of cardinality $|S| \leq l$ there are at least k distinct vertices $\{v_i\}_{i=1}^{k}$ such that $S \subseteq N(v_1)$ for $i \in [1, k]$.

Reference [85] define k-candidate anonymity when k-anonymity is in terms of a structural query Q. For appropriate queries Q, this definition encompases the previous ones.

Definition 6.2 Let Q be a structural query. An anonymized graph satisfies k-candidate anonymity given Q if for all x in V and all y in $cand_Q(x)$

$$C_{Q,x}[y] \leq 1/k$$

where $C_{Q,x}[y]$ is the probability, given Q, of taking candidate y for x. The authors define $C_{Q,x}[y] = 1/|cand_Q(x)|$ for each $y \in cand_Q(x)$ and 0 otherwise.

Each of these definitions for k-anonymity are published together with algorithms to achieve privacy in such terms. The definitions started research lines on efficient algorithms for achieving compliant masked networks. For example, [68] is an algorithm for k-degree anonymity. A survey of techniques for privacy on graphs and social networks can be found in [144].

6.5 Data Protection Procedures for Constrained Data

Values of different variables in a dataset are often related, and dependent. When a database is defined, some of these dependences are expressed by means of constraints between variables. For example, ages are positive, and the number of harvested acres is less than the number of planted acres. These constraints can be stated in the metadata or the schema of the database. When we protect data, these constraints should be taken into account.

In statistical offices it is usual that data is edited before its publication. Data editing [145, 146] (see also Fig. 6.2) studies and develops methods to analyze and correct raw data so that it is compliant with the set of constraints that we have assigned to a database. This is needed for the sake of correctness and so data does satisfy expectations (e.g. no negative ages). Data privacy usually presumes that data is already compliant with the constraints. In statistical disclosure control, it is usual to apply data editing to the original raw data and, in any case, before any perturbation takes place.

Most masking methods ignore the constraints the data are forced to satisfy. This can cause that e.g. a random perturbation of the data invalidates some constraints. For example, adding noise to a data set can cause some ages to be negative or greater than 150 years!, or it can cause some nonsensical combination as e.g. replace the record $(18, woman, pregnant)$ representing a 18 years old pregnant woman by $(2, man, pregnant)$.

The posteriori edition of the masked dataset is always possible but it can cause some additional problems. Consider, for example, a masked data set with information concerning ages and salaries of a set of people. Then, if we add noise in a way that mean is not modified we may have negative ages and salaries. To solve this problem one may consider replacing all salaries and ages by zero. But, this process causes the mean age and the mean salary of the people in the sample to increase.

Data editing can be defined as the process of manipulating raw data to reduce the errors present and improve their quality. Other definitions exist strengthening different aspects of the process. Pierzchala [340] distinguishes data editing as either a validating procedure or a statistical process. In the first case, understanding it as a validating procedure, edition is a "within-record action with the emphasis on detecting inconsistencies, impossibilities, and suspicious situations and correcting them. Examples of validation include: checking to see if the sum of parts adds up to the total, checking that the number of harvested acres is less than or equal to that of planted acres". In the second case, editing as a statistical process, we have that "checks are based on a statistical analysis of respondent data". So, while in the first case we have single record checking, in this second case we may have between-record checking. The detection of outliers belong to this second case.

Given a record that does not satisfy a constraint, it is not easy to know which variable or set of variables are the wrong ones. A common approach used for error localization is the one described in [160]. It looks for the minimum number of variables to be changed in a record so all edit rules are satisfied.

Besides of edit constraints, there is also macro-editing, which consists of checking aggregations of data. This is used when data is aggregated before its publication (e.g. to construct tabular data – following statistical disclosure control jargon – or data summaries).

Fig. 6.2 Definitions of data editing

A few masking methods have been developed that take into account edit constraints. When such constraints are explicitly stated, their goal is to mask the data taking them into account and producing a masked data set compliant with them.

Before discussing such masking methods, we review the constraints usually considered. We classify them according to [148].

6.5.1 Types of Constraints

We distinguish the following types of constraints. These types are not exclusive and some constraints can be seen as belonging to more than one class.

- **Constraints on the possible values**. The values of a given variable are restricted to a predefined set. For example, salary and age have positive values and, possibly, bounded in an interval. For example,

$$\text{EC-PV: } age \in [0, 125]$$

We can generalize this constraint to sets of variables. That is, consider values for a pair (v_1, v_2). Reference [149] consider conditions as $v_1/v_2 \in [l, u]$. In some cases the set of possible values may also be expressed in terms of an equation. For example, if v_1 and v_2 represent the proportion of cases with and without an illness in a hospital, then the values should add up to one. In this latter case, we can represent this constraint as a linear constraint, a type discussed below.
- **Values are restricted to exist in the domain**. We require that the values of a variable belong to a predefined set (as in the constraints on the possible values),

but the values should really exist in the original data set. For example, a variable *age* is required to be in the range [0, 120] but also that the values exist in the population. The set of values has been obtained by other means. When data is masked we add this constraint when we want that masked data contains values in the original data set. It is also useful when data editing is to be used on a file constrained by the data in another file [150, 151] (e.g. when linked files are edited). For example, the edition of a file with data from a school typically should be in agreement with the population data from the same town or neighborhood.

- **One variable governs the possible values of another one**. Given a variable v_1, we have that the values of a variable v_2 are constrained by the values of v_1. For example, we can consider as in [152] the variable *age* governing the variable *married*. It is clear that not all values are acceptable for the variable *married* when $age = 2$. Formally,

EC-GV1: If $age = 2$ THEN *married* $=$ FALSE

Or, in general, we would define a domain for age where married should be false according to e.g. legal constraints.

Reference [152] gives another example represented in the following rule, which links three variables: *age, gross income* and *mean income*.

EC-GV2: IF $age < 17$ THEN *gross income* $<$ *mean income*

Finally, we repeat the example from [146] mentioned above: the number of harvested acres should be less than or equal to that of planted acres. That is,

EC-GV3: *harvested acres* $\leq$ *planted acres*

- **Linear constraints**. In this case we have a numerical (and linear) relationship between some numerical variables. For example, in a data set about economical data the following equation involving variables *net, tax* and *gross* should hold:

EC-LC1: *net* $+$ *tax* $=$ *gross*

- **Non-linear constraints**. This case is similar to the previous one but with a non-linear relationship. The following rule illustrates a non-linear relationship between the variables *applicable VAT Rate, price exc. VAT* and *retail price*.

EC-NLC1: *price exc. VAT* $\cdot$ *(1.00 + applicable VAT Rate)* $=$ *retail price*

Reference [153] gives another example of non-linear relationship based on the variables *wage sum, hours paid for*, and *wage rate*.

EC-NLC2: *wage sum* $=$ *hours paid for* $\cdot$ *wage rate*

- **Other types of constraints**. Constraints not classified in the previous classes are
assigned here. For example, constraints on non-numerical variables (ordinal or
categorical variables), and relationships between several variables that can not be
easily represented by an equation.

The following example, from [148], illustrates edit constraints.

Example 6.3 [148] Table 6.2 represents a file with data from 12 individuals described
in terms of the following 7 variables: *Expenditure at 16%, Expenditure at 7%, Total Expenditure, Hours paid for, Wage rate, Wage sum, Total hours.* We also use
$V_1, \ldots, V_7$ to denote these variables. These variables satisfy three edit constraints.
   The first constraint is a linear one that involves variables V_1, V_2, and V_3. They
satisfy $V_3 = 1.16V_1 + 1.07V_2$. The second one is a multiplicative constraint. Variables V_4, V_5, and V_6 satisfy $V_6 = V_4 * V_5$. The third one is an inequality. Variables
V_4 and V_7 satisfy $V_4 \leq V_7$.

First results for masking data taking into account edit constraints were presented
in [152, 154]. Reference [154] focus on PRAM. The authors propose to use imputation for correcting records that do not satisfy the constraints. The authors also discuss that some of the strategies to decrease disclosure risk can reduce the number of
records not satisfying the constraints. In particular, the authors mention compound-
ing different variables into a single one (and apply PRAM to this new variable).
Reference [152] focus on additive noise and microaggregation. The approach consists of correcting the records that violated the constraints after masking by means of

Table 6.2 A data file satisfying three edit constraint. An additive one involving V_1, V_2, V_3, a
multiplicative one involving V_4, V_5, V_6 and an inequality involving V_4, V_7. Example from [148]

Exp 16% V_1	Exp 7% V_2	Total V_3	Hours paid for V_4	Wage rate V_5	Wage sum V_6	Total hours V_7
15	23	42.01	23	50	1150	37
12	43	59.93	28	70	1960	37
64	229	319.27	12	84	1008	25
12	45	62.07	29	73	2117	30
28	39	74.21	9	30	270	40
71	102	191.5	10	63	630	20
23	64	95.16	9	74	666	10
25	102	138.14	72	30	2160	80
48	230	301.78	26	30	780	35
32	50	90.62	6	45	270	15
90	200	318.4	8	45	360	15
16	100	125.56	34	55	1870	45

linear programming. The approach in [155] is similar. It is based on noise addition. When constraints are not satisfied, a swapping process between records is applied so that records satisfy the constraints. Reference [149] applies a data protection method to the data and then replaces those records that are not compliant with the constraints by values generated using an imputation procedure. The imputation procedure is defined (see e.g. [156] for details) in such a way that the records generated satisfy the contraints. Therefore, all these methods apply a post-masking approach to correct the incorrect records.

Reference [148] focuses on a different approach. It directly defines a masking method so that the constraints are satisfied once the masking method is applied, and there is no need for a post-masking process. The approach is based on microaggregation. To satisfy the constraints, the variables are partitioned into the same set when they jointly take part in the same constraint. Then, the author derives how to compute the cluster center for different types of constraints. For example, for linear constraints cluster representatives can be computed using the arithmetic mean, but this approach does not work for non-linear constraints. For variables satisfying non-linear constraints of the form $V = \Pi V_i^{\alpha_i}$, the cluster centers have to be computed using the geometric mean. Results are proven using functional equations [157]. Reference [158] describes a system that protects a data file with different types of constraints (expressed in Schematron [159]) selecting the appropriate microaggregation masking method for each subset of variables.

6.6 Masking Methods and Big Data

This section summarizes some of the results we have already seen in this chapter, but from the big data perspective. We will classify them according to the distinction we have given in Sect. 3.3.

- **(i) Large volumes**. We have reviewed in the previous sections methods for data of large volumes: both a large number of records and a high dimension. We discussed swapping (Sect. 6.1.1), microaggregation (Sect. 6.1.2), additive noise, PRAM (Sect. 6.1.3), and k-anonymity (Sect. 6.4.1).

  We have discussed methods for a few types of databases that are usually very large, as e.g. time series, location privacy, graphs and social networks, and logs. For unstructured data, see e.g. [160].

  Additive noise is naturally the best option with respect to time, however, it is not usually so good with respect to the trade-off between information loss and disclosure risk. Other methods as microaggregation are a second option for large databases.

- **(ii) Streaming data**. Data protection methods are usually applied to a sliding window, as it is unfeasible to process all data at once. Optimal solutions do not exist, and heuristic methods are applied. As records are delayed by the masking method, an additional constraint is that the maximum delay of a record is small.

There are a few algorithms for masking data streams. Some of them describe algorithms to achieve k-anonymity. See e.g. [161–164]. Others are variants of microaggregation (see e.g., [165]) and rank swapping [28]. For numerical data, [166] is also suitable.

An important aspect when we are masking data by means of building k-anonymous sets is to ensure that these sets really contain at least k different identities. In other words, that we are not including in the set several records from the same identity. This issue is discussed in several works. See e.g. [167,168]. This particular problem in the case of data streams is considered in [164].

- **(iii) Dynamic data**. In Sect. 5.9.3 we have summarized Example 4 in [169] that illustrates the difficulty of ensuring privacy in this context. Two releases of data from a school class are done independently. Only one child was born in February while there are two childs in the other months. Independent releases of $k = 2$ anonymity can lead to disclosure if we anonymize once the February child with the ones in January and once with the ones in March.

Algorithms for protection of dynamic data can be found in [169–173]. Reference [173] focuses on databases with textual documents.

References

1. Carrol, L.: Alice's adventures in wonderland. In project Gutenberg's (1865)
2. Adam, N.R., Wortmann, J.C.: Security-control for statistical databases: a comparative study. ACM Comput. Surv. **21**, 515–556 (1989)
3. Domingo-Ferrer, J., Torra, V.: Disclosure control methods and information loss for microdata. In: Doyle, P., Lane, J.I., Theeuwes, J.J.M., Zayatz, L. (eds.) Confidentiality, Disclosure, and Data Access: Theory and Practical Applications for Statistical Agencies, North-Holland, pp. 91–110 (2001)
4. Willenborg, L., de Waal, T.: Elements of Statistical Disclosure Control. Lecture Notes in Statistics. Springer, New York (2001)
5. Hundepool, A., Domingo-Ferrer, J., Franconi, L., Giessing, S., Nordholt, E.S., Spicer, K., de Wolf, P.-P.: Statistical Disclosure Control. Wiley, New York (2012)
6. Duncan, G.T., Elliot, M., Salazar, J.J.: Statistical Confidentiality. Springer, New York (2011)
7. Felsö, F., Theeuwes, J., Wagner, G.: Disclosure limitation in use: results of a survey. In: Doyle, P., Lane, J.I., Theeuwes, J.J.M., Zayatz, L. (eds.) Confidentiality, Disclosure, and Data Access: Theory and Practical Applications for Statistical Agencies, North-Holland, pp. 17–42 (2001)
8. Templ, M.: Statistical disclosure control for microdata using the R-Package sdcMicro. Trans. Data Priv. **1**, 67–85 (2008)
9. Hundepool, A., van de Wetering, A., Ramaswamy, R., Franconi, L., Capobianchi, C., de Wolf, P.-P., Domingo-Ferrer, J., Torra, V., Brand, R., Giessing, S.: μ-ARGUS version 3.2 Software and User's Manual, Voorburg NL, Statistics Netherlands, February 2003. http://neon.vb.cbs.nl/casc. Version 4.0 published on may 2005
10. Dalenius, T., Reiss, S.P.: Data-swapping—a technique for disclosure control. In: Proceedings of ASA Section on Survey Research Methods, pp. 191–194 (1978)
11. Dalenius, T., Reiss, S.P.: Data-swapping: a technique for disclosure control. J. Stat. Plan. Infer. **6**, 73–85 (1982)

12. Reiss, S.P.: Practical data-swapping: the first steps. In: Proceedings of 1980 Symposium on Security and Privacy, pp. 38–45 (1980)
13. Reiss, S.P.: Practical data-swapping: the first steps. ACM Trans. Database Syst. **9**(1), 20–37 (1984)
14. Fienberg, S.E., McIntyre, J.: Data swapping: variations on a theme by Dalenius and Reiss. In: Proceedings of the PSD 2004. LNCS, vol. 3050, pp. 14–29 (2004)
15. Greenberg, B.: Rank Swapping for Masking Ordinal Microdata, US Bureau of the Census (1987, unpublished manuscript)
    N=Y
16. Moore, R.: Controlled data swapping techniques for masking public use microdata sets, U.S. Bureau of the Census (1996, unpublished manuscript)
17. Domingo-Ferrer, J., Torra, V.: A quantitative comparison of disclosure control methods for microdata. In: Doyle, P., Lane, J.I., Theeuwes, J.J.M., Zayatz, L. (eds.) Confidentiality, Disclosure and Data Access: Theory and Practical Applications for Statistical Agencies, North-Holland, pp. 111–134 (2001)
18. Torra, V.: Microaggregation for categorical variables: a median based approach. In: Proceedings of PSD 2004. LNCS, vol. 3050, pp. 162–174 (2004)
19. Carlson, M., Salabasis, M.: A data swapping technique using ranks: a method for disclosure control. Res. Official Stat. **5**(2), 35–64 (2002)
20. Takemura, A.: Local recoding and record swapping by maximum weight matching for disclosure control of microdata sets. J. Official Stat. **18**, 275–289 (1999). Local recoding by maximum weight matching for disclosure control of microdata sets (2002)
21. Muralidhar, K., Sarathy, R.: Data shuffling a new masking approach for numerical data. Manag. Sci. **52**(5), 658–670 (2006)
22. Nin, J., Herranz, J., Torra, V.: Rethinking rank swapping to decrease disclosure risk. Data Knowl. Eng. **64**(1), 346–364 (2007)
23. Muralidhar, K., Domingo-Ferrer, J.: Rank-based record linkage for re-identification risk assessment. In: Proceedings of PSD (2016)
24. Torra, V.: Rank swapping for partial orders and continuous variables. In: Proceedings of ARES 2009, pp. 888–893 (2009)
25. Lasko, T.A., Vinterbo, S.A.: Spectral anonymization of data. IEEE Trans. Knowl. Data Eng. **22**(3), 437–446 (2010)
26. Lasko, T.A.: Spectral anonymization of data. Ph.D. dissertation, MIT (2007)
27. Gidófalvi, G.: Spatio-temporal data mining for location-based services. Ph.D. dissertation (2007)
28. Navarro-Arribas, G., Torra, V.: Rank swapping for stream data. In: Proceedings of MDAI 2014. LNCS, vol. 8825, pp. 217–226 (2014)
29. Defays, D., Nanopoulos, P.: Panels of enterprises and confidentiality: the small aggregates method. Proceedings of 92 Symposium on Design and Analysis of Longitudinal Surveys, Statistics Canada, pp. 195–204 (1993)
30. Hansen, S., Mukherjee, S.: A polynomial algorithm for optimal univariate microaggregation. IEEE Trans. Knowl. Data Eng. **15**(4), 1043–1044 (2003)
31. Oganian, A., Domingo-Ferrer, J.: On the complexity of optimal microaggregation for statistical disclosure control. Stat. J. United Nations Econ. Comm. Europe **18**(4), 345–354 (2000)
32. Aggarwal, C.: On k-anonymity and the curse of dimensionality. In: Proceedings of the 31st International Conference on Very Large Databases, pp. 901–909 (2005)
33. Nin, J., Herranz, J., Torra, V.: How to group attributes in multivariate microaggregation. Int. J. Unc. Fuzz. Knowl. Based Syst. **16**(1), 121–138 (2008)
34. Balasch-Masoliver, J., Muntés-Mulero, V., Nin, J.: Using genetic algorithms for attribute grouping in multivariate microaggregation. Intell. Data Anal. **18**, 819–836 (2014)
35. Falkenauer, E.: Genetic Algorithms and Grouping. Wiley, Chichester (1998)

36. Sun, X., Wang, H., Li, J.: Microdata protection through approximate microaggregation. In: Proceedings of CRPIT (2009)
37. Chow, C., Liu, C.: Approximating discrete probability distributions with dependence trees. IEEE Trans. Inf. Theor. **14**(3), 462–467 (1968)
38. Oommen, B.J., Fayyoumi, E.: On utilizing dependence-based information to enhance microaggregation for secure statistical databases. Pattern Anal. Appl. **16**, 99–116 (2013)
39. Domingo-Ferrer, J., Mateo-Sanz, J.M.: Practical data-oriented microaggregation for statistical disclosure control. IEEE Trans. Knowl. Data Eng. **14**(1), 189–201 (2002)
40. Domingo-Ferrer, J., Torra, V.: Ordinal, continuous and heterogeneous k-anonymity through microaggregation. Data Mining Knowl. Discov. **11**(2), 195–212 (2005)
41. Sande, G.: Exact and approximate methods for data directed microaggregation in one or more dimensions. Int. J. Unc. Fuzz. Knowl. Based Syst. **10**(5), 459–476 (2002)
42. Laszlo, M., Mukherjee, S.: Minimum spanning tree partitioning algorithm for microaggregation. IEEE Trans. Knowl. Data Eng. **17**(7), 902–911 (2005)
43. Lin, J.-L., Wen, T.-H., Hsieh, J.-C., Chang, P.-C.: Density-based microaggregation for statistical disclosure control. Expert Syst. Appl. **37**, 3256–3263 (2010)
44. Chiu, S.L.: A cluster estimation method with extension to fuzzy model identification. In: Proceedings of IEEE Fuzzy Systems (1994)
45. Chang, C.-C., Li, Y.-C., Huang, W.-H.: TFRP: an efficient microaggregation algorithm for statistical disclosure control. J. Syst. Softw. **80**, 1866–1878 (2007)
46. Torra, V., Narukawa, Y.: Modeling Decisions: Information Fusion and Aggregation Operators. Springer, Berlin (2007)
47. Stokes, K., Torra, V.: n-Confusion: a generalization of k-anonymity. In: Proceedings of Fifth International Workshop on Privacy and Anonymity on Information Society, PAIS (2012)
48. Stokes, K., Torra, V.: Blow-up microaggregation: satisfying variance, manuscript (2011)
49. Domingo-Ferrer, J., González-Nicolás, U.: Hybrid microdata using microaggregation. Inf. Sci. **180**, 2834–2844 (2010)
50. Nin, J., Herranz, J., Torra, V.: On the disclosure risk of multivariate microaggregation. Data Knowl. Eng. **67**(3), 399–412 (2008)
51. Winkler, W.E.: Single ranking micro-aggregation and re-identification, Statistical Research Division report RR 2002/08 (2002)
52. Torra, V., Miyamoto, S.: Evaluating fuzzy clustering algorithms for microdata protection. In: Proceedings of PSD 2004. LNCS, vol. 3050, pp. 175–186 (2004)
53. Nin, J., Torra, V.: Analysis of the univariate microaggregation disclosure risk. New Gener. Comput. **27**, 177–194 (2009)
54. Domingo-Ferrer, J., Torra, V.: Towards fuzzy c-means based microaggregation. In: Grzegorzewski, P., Hryniewicz, O., Gil, M.A. (eds.) Soft Methods in Probability and Statistics, pp. 289–294 (2002)
55. Domingo-Ferrer, J., Torra, V.: Fuzzy microaggregation for microdata protection. J. Adv. Comput. Intell. Intell. Inf. **7**(2), 53–159 (2003)
56. Torra, V.: A fuzzy microaggregation algorithm using fuzzy c-means. In: Proceedings of CCIA, pp. 214–223 (2015)
57. Torra, V.: Fuzzy microaggregation for the transparency principle. J. Appl. Logics (2017, in press)
58. Muntés-Mulero, V., Nin, J.: Privacy and anonymization for very large datasets. In: Proceedings of 18th ACM IKM, pp. 2117–2118 (2009)
59. Solé, M., Muntés-Mulero, V., Nin, J.: Efficient microaggregation techniques for large numerical data volumes. Int. J. of Inf. Secur. **11**(4), 253–267 (2012)
60. Mortazavi, R., Jalili, S.: Fast data-oriented microaggregation algorithm for large numerical datasets. Knowl. Based Syst. **67**, 192–205 (2014)

61. Salari, M., Jalili, S., Mortazavi, R.: TBM, a transformation based method for microaggregation of large volume mixed data. Data Mining Knowl. Disc. **31**(1), 65–91 (2016). doi:10.1007/s10618-016-0457-y

62. Abril, D., Navarro-Arribas, G., Torra, V.: Spherical microaggregation: anonymizing sparse vector spaces. Comput. Secur. **49**, 28–44 (2015)

63. Nin, J., Torra, V.: Extending microaggregation procedures for time series protection. LNCS, vol. 4259, pp. 899–908 (2006)

64. Abul, O., Bonchi, F., Nanni, M.: Never walk alone: uncertainty for anonymity in moving objects databases. In: Proceedings of 24th ICDE 2008, pp. 376–385 (2008)

65. Abul, O., Bonchi, F., Nanni, M.: Anonymization of moving objects databases by clustering and perturbation. Inf. Sci. **35**, 884–910 (2010)

66. Domingo-Ferrer, J., Trujillo-Rasúa, R.: Microaggregation- and permutation-based anonymization of movement data. Inf. Sci. **208**, 55–80 (2012)

67. Ferreira Torres, C., Trujillo-Rasua, R.: The Fréchet/Manhattan distance and the trajectory anonymisation problem. In: Proceedings of DBSec 2016. LNCS, vol. 9766, pp. 19–34 (2016)

68. Casas-Roma, J., Herrera-Joancomartí, J., Torra, V.: An algorithm for k-degree anonymity on large networks. In: Proceedings of 2013 IEEE/ACM ASONAM (2013)

69. Salas, J., Torra, V.: Graphic sequences, distances and k-degree anonymity. Discrete Appl. Math. **188**, 25–31 (2015)

70. Hay, M., Miklau, G., Jensen, D., Towsley, D.F., Li, C.: Resisting structural reidentification in anonymized social networks. J. VLDB **19**, 797–823 (2010)

71. Prost, F., Yoon, J.: Parallel clustering of graphs for anonymization and recommender systems. Arxiv (2016)

72. Navarro-Arribas, G., Torra, V.: Tree-based microaggregation for the anonymization of search logs. In: Proceedings of 2009 IEEE/WIC/ACM WI 2009, pp. 155–158 (2009)

73. Erola, A., Castellà-Roca, J., Navarro-Arribas, G., Torra, V.: Semantic microaggregation for the anonymization of query logs. In: Proceedings of PSD 2010. LNCS, vol. 6344, pp. 127–137 (2010)

74. Batet, M., Erola, A., Sánchez, D., Castellà-Roca, J.: Utility preserving query log anonymization via semantic microaggregation. Inf. Sci. **242**, 49–63 (2013)

75. Liu, J., Wang, K.: Anonymizing bag-valued sparse data by semantic similarity-based clustering. Knowl. Inf. Syst. **35**, 435–461 (2013)

76. Spruill, N.L.: The confidentiality and analytic usefulness of masked business microdata. In: Proceedings of the Section on Survery Research Methods, vol. 1983, pp. 602–610. American Statistical Association (1983)

77. Fuller, W.A.: Masking procedures for microdata disclosure limitation. J. Official Stat. **9**, 383–406 (1993)

78. Brand, R.: Microdata protection through noise addition. In: Domingo-Ferrer, J. (ed.) Proceedings of Inference Control in Statistical Databases. LNCS, vol. 2316, pp. 97–116 (2002)

79. Domingo-Ferrer, J., Sebe, F., Castella-Roca, J.: On the security of noise addition for privacy in statistical databases. In: Proceedings of PSD 2004. LNCS, vol. 3050, pp. 149–161 (2004)

80. Agrawal, R., Srikant, R.: Privacy preserving data mining. In: Proceedings of of the ACM SIGMOD Conference on Management of Data, pp. 439–450 (2000)

81. Huang, Z., Du, W., Chen, B.: Deriving private information from randomized data. In: Proceedings of SIGMOD 2005, pp. 37–48 (2005)

82. Kim, J., Winkler, W.: Multiplicative noise for masking continuous data. U.S. Bureau of the Census, RR2003/01 (2003)

83. Liu, K., Kargupta, H., Ryan, J.: Random projection based multiplicative data perturbation for privacy preserving data mining. IEEE Trans. Knowl. Data Eng. **18**(1), 92–106 (2006)

84. Rodriguez-Garcia, M., Batet, M., Sanchez, D.: Semantic noise: privacy-protection of nominal microdata through uncorrelated noise addition. In: Proceedings of 27th ICTAI (2015)

85. Hay, M., Miklau, G., Jensen, D., Weis, P., Srivastava, S.: Anonymizing Social Networks. Computer science Department Faculty publication series, p. 180 (2007)
86. Casas-Roma, J., Herrera-Joancomartí, J., Torra, V.: Comparing random-based and k-anonymity-based algorithms for graph anonymization. In: Proceedings of MDAI 2012. LNCS, vol. 7647, pp. 197–209 (2012)
87. Gouweleeuw, J.M., Kooiman, P., Willenborg, L.C.R.J., De Wolf, P.-P.: Post randomisation for statistical disclosure control: theory and implementation. J. Official Stat. **14**(4), 463–478 (1998). Also as Research Paper No. 9731. Statistics Netherlands, Voorburg (1997)
88. De Wolf, P.P., Van Gelder, I.: An empirical evaluation of PRAM, Discussion paper 04012. Statistics Netherlands, Voorburg/Heerlen (2004)
89. Gross, B., Guiblin, P., Merrett, K.: Implementing the post randomisation method to the individual sample of anonymised records (SAR) from the 2001 census. In: The Samples of Anonymised Records, An Open Meeting on the Samples of Anonymised Records from the 2001 Census (2004)
90. Marés, J., Torra, V.: An evolutionary algorithm to enhance multivariate post-randomization method (PRAM) protections. Inf. Sci. **278**, 344–356 (2014)
91. Marés, J., Shlomo, N.: Data privacy using an evolutionary algorithm for invariant PRAM matrices. Comput. Stat. Data Anal. **79**, 1–13 (2014)
92. Marés, J., Torra, V., Shlomo, N.: Optimisation-based study of data privacy by using PRAM. In: Navarro-Arribas G., Torra, V. (eds.) Advanced Research in Data Privacy, pp. 83–108. Springer (2015)
93. Jiménez, J., Torra, V.: Utility and risk of JPEG-based continuous microdata protection methods. In: ARES, Proceedings of International Conference on Availability, Reliability and Security, pp. 929–934 (2009)
94. Jiménez, J., Navarro-Arribas, G., Torra, V.: JPEG-based microdata protection. In: Proceedings of PSD 2014. LNCS, vol. 8744, pp. 117–129 (2014)
95. Parker, J.R.: Practical Computer Vision Using C. Wiley, New York (1994)
96. Bapna, S., Gangopadhyay, A.: A wavelet-based approach to preserve privacy for classification mining. Decis. Sci. **37**(4), 623–642 (2006)
97. Liu, L., Wang, J., Lin, Z., Zhang, J.: Wavelet-based data distortion for privacy-preserving collaborative analysis. Technical report N. 482-07, Department of Computer Science, University of Kentucky (2007)
98. Liu, L., Wang, J., Zhang, J.: Wavelet-based data perturbation for simultaneous privacy-preserving and statistics-preserving. In: IEEE ICDM Workshops (2008)
99. Hajian, S., Azgomi, M.A.: A privacy preserving clustering technique using Haar wavelet transform and scaling data perturbation. IEEE (2008)
100. Hajian, S., Azgomi, M.A.: On the use of Haar wavelet transform and scaling data perturbation for privacy preserving clustering of large datasets. Int. J. Wavelets Multiresolut. Inf. Process. **9**(6), 867 (2011)
101. Mukherjee, S., Chen, Z., Gangopadhyay, A.: A privacy-preserving technique for Euclidean distance-based mining algorithms using Fourier-related transforms. VLDB J. **15**, 293–315 (2006)
102. Xu, S., Zhang, J., Han, D., Wang, J.: Data distortion for privacy protection in a terrorist analysis system. In: Proceedings of IEEE ICISI (2005)
103. Xu, S., Zhang, J., Han, D., Wang, J.: Singular value decomposition based data distortion strategy for privacy protection. Knowl. Inf. Syst. **10**(3), 383–397 (2006)
104. Wang, J., Zhong, W.J., Zhang, J.: NNMF-based factorization techniques for high-accuracy privacy protection on non-negative-valued datasets. In: Proceedings of PADM (2006)
105. Xiao, X., Wang, G., Gehrke, J.: Differential privacy via wavelet transforms. IEEE Trans. Knowl. Data Eng. **23**(8), 1200–1214 (2009)
106. Cochran, W.G.: Sampling Techniques, 3rd edn. Wiley, New York (1977)

107. Lohr, S.: Sampling: Design and Analysis. Duxbury, Pacific Grove (1999)
108. LeFevre, K., DeWitt, D.J., Ramakrishnan, R.: Incognito: efficient full-domain k-anonymity. In: SIGMOD 2005 (2005)
109. LeFevre, K., DeWitt, D.J., Ramakrishnan, R.: Multidimensional k-anonymity. Technical report 1521, University of Wisconsin (2005)
110. Samarati, P., Sweeney, L.: Protecting privacy when disclosing information: k-anonymity and its enforcement through generalization and suppression. SRI International Technical report (1998)
111. Sweeney, L.: k-anonymity: a model for protecting privacy. Int. J. Unc. Fuzz Knowl. Based Syst. **10**(5), 557–570 (2002)
112. Sweeney, L.: Datafly: a system for providing anonymity in medical data. In: Proceedings of IFIP TC11 WG11.3 11th International Conference on Database security XI: Status and Prospects, pp. 356–381 (1998)
113. Templ, M., Meindl, B.: Methods and tools for the generation of synthetic populations. In: A brief review, PSD (2014)
114. Münnich, R., Schürle, J.: On the simulation of complex universes in the case of applying the German Microcensus. DACSEIS research paper series no. 4. University of Tübingen (2003)
115. Münnich, R., Schürle, J., Bihler, W., Boonstra, H.-J., Knotterus, P., Nieuwenbroek, N., Haslinger, A., Laaksoner, S., Eckmair, D., Quatember, A., Wagner, H., Renfer, J.-P., Oetliker, U., Wiegert, R.: Monte Carlo simulation study of European surveys. DACSEIS Deliverables D3.1 and D3.2. University of Tübingen (2003)
116. Barthelemy, J., Toint, P.L.: Synthetic population generation without a sample. Transp. Sci. **47**(2), 266–279 (2013)
117. Little, R.J.A.: Statistical analysis of masked data. J. official Stat. **9**(2), 407–426 (1993)
118. Rubin, D.B.: Discussion: statistical disclosure limitation. J. official Stat. **9**(2), 461–468 (1993)
119. Torra, V., Abowd, J.M., Domingo-Ferrer, J.: Using Mahalanobis distance-based record linkage for disclosure risk assessment. LNCS, vol. 4302, pp. 233–242 (2006)
120. Reiter, J.P., Drechsler, J.: Releasing multiply-imputed synthetic data generated in two stages to protect confidentiality. Statistica Sinica **20**, 405–421 (2010)
121. Reiter, J.P.: Releasing multiply-imputed, synthetic public use microdata: an illustration and empirical study. J. Roy. Statist. Soc. Ser. A **168**, 185–205 (2005)
122. Drechsler, J.: Synthetic Datasets for Statistical Disclosure Control: Theory and Implementation. Springer, New York (2011)
123. Drechsler, J., Bender, S., Rässler, S.: Comparing fully and partially synthetic datasets for statistical disclosure control in the German IAB Establishment Panel. Trans. Data Priv. **1**, 105–130 (2008)
124. Burridge, J.: Information preserving statistical obfuscation. Stat. Comput. **13**, 321–327 (2003)
125. Muralidhar, K., Sarathy, R.: An enhanced data perturbation approach for small data sets. Decis. Sci. **36**(3), 513–529 (2005)
126. Dandekar, R.A., Cohen, M., Kirkendall, N.: Applicability of Latin hypercube sampling technique to create multivariate synthetic microdata. In: Proceedings of ETK-NTTS, pp. 839–847 (2001)
127. Drechsler, J., Reiter, J.P.: An empirical evaluation of easily implemented, nonparametric methods for generating synthetic datasets. Comput. Stat. Data Anal. **55**, 3232–3243 (2011)
128. Meyerson, A., Williams, R.: On the complexity of optimal k-anonymity. In: Proceedings of 23rd ACM-SIGMOD-SIGACT-SIGART Symposium on the Principles of Database Systems, pp. 223–228 (2004)
129. Aggarwal, G., Feder, T., Kenthapadi, K., Motwani, R., Panigrahy, R., Thomas, D., Zhu, A.: Anonymizing tables. In: Proceedings of 10th International Conference on Database Theory (ICDT05), pp. 246–258 (2005)

130. Sun, X., Wang, H., Li, J.: On the complexity of restricted k-anonymity problem. In: Proceedings of 10th Asia Pacific Web Conference (APWEB2008). LNCS, vol. 4976, pp. 287–296 (2008)

131. Friedman, J., Bentley, J., Finkel, R.: An algorithm for finding best matchings in logarithmic expected time. ACM Trans. Math. Softw. **3**(3), 209–226 (1977)

132. Byun, J.-W., Kamra, A., Bertino, E., Li, N.: Efficient k-anonymization using clustering techniques. In: Proceedings of DASFAA (2007)

133. Samarati, P.: Protecting respondents' identities in microdata release. IEEE Trans. Knowl. Data Eng. **13**(6), 1010–1027 (2001)

134. Sun, X., Li, M., Wang, H., Plank, A.: An efficient hash-based algorithm for minimal k-anonymity. In: Proceedings of ACSC (2008)

135. Russom, Y.K.: Privacy preserving for big data analysis. Master's thesis, University of Stavanger (2013)

136. Deng, Z., Zhu, X., Cheng, D., Zong, M., Zhang, S.: Efficient kNN classification algorithm for big data. Neurocomputing **195**, 143–148 (2016)

137. Nergiz, M.E., Atzori, M., Saygın, Y.: Towards trajectory anonymization: a generalization-based approach. In: Proceedings of SIGSPATIAL ACM GIS International Workshop on Security and Privacy in GIS and LBS (2008)

138. Monreale, A., Andrienko, G., Andrienko, N., Giannotti, F., Pedreschi, D., Rinzivillo, S., Wrobel, S.: Movement data anonymity through generalization. Trans. Data Priv. **3**, 91–121 (2010)

139. Shokri, R., Troncoso, C., Diaz, C., Freudiger, J., Hubaux, J.-P.: Unraveling an old cloak: k-anonymity for location privacy. In: Proceedings of WPES (2010)

140. Campan, A., Truta, T.M.: Data and structural k-anonymity in social networks. In: Proceedings of PinkDD. LNCS, vol. 5456, pp. 33–54 (2008)

141. Liu, K., Terzi, E.: Towards identity anonymization on graphs. In: Proceedings of SIGMOD (2008)

142. Zhou, B., Pei, J.: Preserving privacy in social networks against neighborhood attacks. In: Proceedings of ICDE (2008)

143. Stokes, K., Torra, V.: Reidentification and k-anonymity: a model for disclosure risk in graphs. Soft Comput. **16**(10), 1657–1670 (2012)

144. Casas-Roma, J., Herrera-Joancomartí, J., Torra, V.: A survey of graph-modification techniques for privacy-preserving on networks. Artif. Intell. Rev. **47**(3), 341–366 (2017). doi:10.1007/s10462-016-9484-8

145. Granquist, L.: The new view on editing. Int. Stat. Rev. **65**(3), 381–387 (1997)

146. Pierzchala, M.: A review of the state of the art in automated data editing and imputation. In: Statistical Data Editing Conference of European Statisticians Statistical Standards and Studies United Nations Statistical Commission and Economic Commission for Europe, vol. 1, no. 44, pp. 10–40 (1994)

147. Fellegi, I.P., Holt, D.: A systematic approach to automatic edit and imputation. J. Am. Stat. Assoc. **71**, 17–35 (1976)

148. Torra, V.: Constrained microaggregation: adding constraints for data editing. Trans. Data Priv. **1**(2), 86–104 (2008)

149. Kim, H.J., Karr, A.F., Reiter, J.P.: Statistical disclosure limitation in the presence of edit rules. J. official Stat. **31**(1), 121–138 (2015)

150. Blum, O.: Evaluation of editing and imputations supported by administrative records. In: Conference of European Statisticians, WP7 (2005)

151. Shlomo, N.: Making use of alternate data sources. In: Statistical Data Editing: Impact on data quality, United Nations Statistical Commission and Economic Commission for Europe, vol. 3, p. 301 (2006)

152. Shlomo, N., De Waal, T.: Protection of micro-data subject to edit constraints against statistical disclosure. J. Official Stat. **24**(2), 229–253 (2008)

153. Gasemyr, S.: Editing and imputation for the creation of a linked micro file from base registers and other administrative data. In: Conference of European Statisticians, WP8 (2005)

154. Shlomo, N., De Waal, T.: Preserving edits when perturbing microdata for statistical disclosure control. In: Conference of European Statisticians, WP11 (2005)

155. Cano, I., Torra, V.: Edit constraints on microaggregation and additive noise. LNCS, vol. 6549, pp. 1–14 (2011)

156. Kim, H.J., Reiter, J.P., Wang, Q., Cox, L.H., Karr, A.F.: Multiple imputation of missing or faulty values under linear constraints. J. Bus. Econ. Stat. **32**(3), 375–386 (2014)

157. Aczél, J.: A Short Course on Functional Equations. D. Reidel Publishing Company (Kluwer Academic Publishers Group), Dordrecht (1987)

158. Cano, I., Navarro-Arribas, G., Torra, V.: A new framework to automate constrained microaggregation. In: Proceedings of PAVLAD Workshop in CIKM (2009)

159. Schematron ISO/IEC: Information technology—Document Schema Definition Language (DSDL)—Part 3: Rule-based validation—Schematron. ISO/IEC 19757-3:2006 Standard JTC1/SC34 (2006)

160. Willemsen, M.: Anonymizing unstructured data to prevent privacy leaks during data mining. In: Proceedings of 25th Twenty Student Conference on IT (2016)

161. Wang, W., Li, J., Ai, C., Li, Y.: Privacy protection on sliding window of data streams. In: Proceedings of ICCC, pp. 213–221 (2007)

162. Li, J., Ooi, B. C., Wang, W.: Anonymizing streaming data for privacy protection. In: Proceedings of 24th ICDE 2008, pp. 1367–1369 (2008)

163. Cao, J., Carminati, B., Ferrari, E., Tan, K.-L.: Castle: a delay-constrained scheme for ks-anonymizing data streams. In: Proceedings of 24th ICDE, pp. 1376–1378 (2008)

164. Zhou, B., Han, Y., Pei, J., Jiang, B., Tao, Y., Jia, Y.: Continuous privacy preserving publishing of data streams. In: Proceedings 12th International Conference on EDBT, pp. 648–659 (2009)

165. Zakerzadeh, H., Osborn, S.L.: FAANST: Fast Anonymizing Algorithm for Numerical Streaming DaTa. In: Proceedings of DPM and SETOP, pp. 36–50 (2010)

166. Li, F., Sun, J., Papadimitriou, S., Mihaila, G.A., Stanoi, I.: Hiding in the crowd: privacy preservation on evolving streams through correlation tracking. In: Proceedings of IEEE 23rd ICDE 2007, pp. 686–695 (2007)

167. De Capitani di Vimercati, S., Foresti, S., Livraga, G., Samarati, P.: Data privacy: definitions and techniques. Int. J. Unc. Fuzz. Knowl. Based Syst. **20**(6), 793–817 (2012)

168. Kifer, D., Machanavajjhala, A.: No free lunch in data privacy. In: Proceedings of SIGMOD (2011)

169. Stokes, K., Torra, V.: Multiple releases of k-anonymous data sets and k-anonymous relational databases. Int. J. Unc. Fuzz. Knowl. Based Syst. **20**(6), 839–854 (2012)

170. Pei, J., Xu, J., Wang, Z., Wang, W., Wang, K.: Maintaining k-anonymity against incremental updates. In: Proceedings of SSDBM (2007)

171. Truta, T.M., Campan, A.: K-anonymization incremental maintenance and optimization techniques. In: Proceedings of ACM SAC, pp. 380–387 (2007)

172. Nergiz, M.E., Clifton, C., Nergiz, A.E.: Multirelational k-anonymity. IEEE Trans. Knowl. Data Eng. **21**(8), 1104–1117 (2009)

173. Navarro-Arribas, G., Abril, D., Torra, V.: Dynamic anonymous index for confidential data. In: Proceedings of 8th DPM and SETOP, pp. 362–368 (2013)

Information Loss: Evaluation and Measures

7

> Farfar, får får får?
> Nej, får får inte får,
> får får vattenäpple.
>
> Privacy-preserving Swedish proverb.

Data protection methods introduce distortion to the data. This causes that analyses on the masked data are different to the same analyses performed on the original data. We say that some information is lost in this distortion process.

The literature presents different ways to evaluate this loss. Some just display in a single framework (a plot or a table) the results of the two analyses (original vs. masked data), while others compute an aggregated measure (an information loss measure comparing the two analyses).

In this chapter we discuss first (Sect. 7.1) the difference between generic and specific information loss and (Sect. 7.2) we formalize measures to quantify information loss. Then (Sects. 7.3, 7.4, and 7.5) we give an overview of information loss measures. First, generic measures, then specific, and finally measures for big data.

7.1 Generic Versus Specific Information Loss

When we know which type of analysis the data scientist will perform on the data, the analysis of the distortion of a particular protection procedure can be done in detail. That is, measures can be developed, and protection procedures can be compared and ranked using such measures. *Specific information loss measures* are the indices that permits us to quantify such distortion.

© Springer International Publishing AG 2017
V. Torra, *Data Privacy: Foundations, New Developments and the Big Data Challenge*, Studies in Big Data 28,
DOI 10.1007/978-3-319-57358-8_7

Nevertheless, when the type of analysis to be performed is not known, only generic indices can be computed. *Generic information loss measures* are the indices to be applied in this case. They have been defined to evaluate the utility of the protected data but not for a specific application but for *any* of them. They are usually defined in terms of an aggregation of a few measures, and are based on statistical properties of the data. As these indices aggregate components, it might be the case that a protection procedure with a good *average* performance behaves badly in a specific analysis.

Generic information loss has been evaluated considering the values of the records (see e.g. [1–5]), ranks of values [6], summary statistics [7,8] (means, variance, covariances), regression coefficients [7,9–12], n-order statistics [13], subgroup analysis (as e.g. means for some combinations [7]), coverage rates for 95% confidence intervals [5,11,12]. We discuss some generic information loss measures in Sect. 7.3.

Specific information loss have been defined for e.g. clustering (k-means [14,15]), classification [14,16–18] (including e.g. k-nearest neighbor, Naïve Bayes, (linear) support vector machines, decision trees), and regression [10] (comparison of estimates). We discuss some specific information loss measures in Sect. 7.4.

7.2   Information Loss Measures

Information loss measures have been defined to quantify information loss. From a general perspective, and for a given data analysis, information loss corresponds to the divergence between the results of the analysis on the original data and the results of the same analysis on the perturbated data. Naturally, the larger the divergence, the larger the information loss.

Definition 7.1 Let X be the original data set on the domain D, and let X' be a protected version of the same data set. Then, for a given data analysis that returns results in a certain domain D' (i.e., $f : D \rightarrow D'$), the information loss of f for data sets X and X' is defined by

$$IL_f(X, X') = divergence(f(X), f(X')), \tag{7.1}$$

where *divergence* is a way to compare two elements of D'.

An analysis of this function seems to point out that it should satisfy for all $X, Y \in D'$ the following axioms:

- $divergence(X, X) = 0$
- $divergence(X, Y) \geq 0$
- $divergence(X, Y) = divergence(Y, X)$

So, *divergence* is a semimetric on D' instead of a metric or distance (because we do not require to satisfy the triangle inequality). Naturally, any metric or distance[1] function on D' will also be acceptable for computing a *divergence*.

In some circumstances, the condition of symmetry can be removed. Let us consider the case of a function f that distinguishes some objects from X. For example, $f(X)$ selects the sensors in X that malfunction. Consider that having a sensor not working properly can cause major damage while informing that a valid sensor is malfunctioning is not so relevant (because double testing a sensor is easy and replacing it by a new one has a negligible cost). Then, our goal is to avoid missing any malfunctioning sensors in X. In this context we can use the following divergence measure that is not symmetric:

$$divergence(X, Y) = |malfunctioning(X) \backslash malfunctioning(Y)|$$

Note that this type of measure focuses on false positives, but ignores false negatives.

In addition to the previous axioms, for the sake of commensurability, we usually require the function *divergence* to be bound to compare protection methods. E.g., $0 \leq divergence(X, Y) \leq 1$.

This definition pressumes a particular data use f. Then, different data uses f imply different information loss measures.

Note also that different types of data D usually imply different information loss measures. This is due to the fact that different D usually imply different functions f. However, it is important to underline that this is not always the case. Observe that, for example, clustering algorithms can be applied to different domains D leading in all cases to a partition of the elements in D. In this case, similar information loss measures can be developed for different data types.

For illustration, we consider now two examples. One for numerical databases and another for search logs.

Example 7.1 Let X be a one-column matrix of numerical data, and X' a protected version of X. Let $f(X) = \bar{X}$ be the average of the matrix X. Then, we can define information loss by

$$IL_{\bar{X}}(X, X') = ||\bar{X} - \bar{X}'||.$$

Example 7.2 Let X be a search log database, and X' be the protected version of X. Let $f(X)$ be the list of ten more frequent queries in X. Then, we can define information loss by

$$IL_{10+}(X, X') = 1 - \frac{|f(X) \cap f(X')|}{|f(X) \cup f(X')}$$

---

[1]Recall that we have discussed distances and metrics, as well as their properties in Sects. 5.4.7 and 5.6.1.

In the first example, $f(X) = \bar{X}$ and $divergence(x, y) = ||x - y||$. In the second example, f selects the ten most frequent logs in the set and

$$divergence(x, y) = 1 - \frac{|x \cap y|}{|x \cup y|},$$

for sets x and y. In the first example D' is the set of real numbers and in the second example D' is a set of logs.

As the two examples above illustrate, different data imply different analyses, and these analyses lead to quite different information loss measures. In addition, for any set, a large number of different data uses can also be conceived.

In the next sections, we discuss some of the existing measures. We start with some generic measures and then we focus on specific measures for classification and clustering.

7.3 Generic Information Loss Measures

References [2, 19] were some of the first papers on defining generic information loss measures. They were defined for numerical and categorical microdata. We review these measures below together with some extensions.

7.3.1 Numerical Data

Information loss measures for numerical data are based on some statistics computed from the data. Some works compute them for the whole domain and others compute them for the whole domain but also for some subdomains (e.g., for subsets defined in terms of a few variables). See e.g. [7, 20] for the latter approach.

There are a few works that compare means, variance, k-th central moments, and covariance. References [2, 19] proposed a definition that from the point of view of Definition 7.1 can be seen as: (i) computation of some matrices from the data sets (as e.g. covariance matrices) which correspond to the computation of function f, and (ii) computation of divergence between these matrices.

Three alternative definitions for divergence were considered: mean square error, mean absolute error, and mean variation. Mean square error is defined as the sum of squared componentwise differences between pairs of matrices, divided by the number of cells in either matrix. Mean absolute error is defined as the sum of absolute componentwise difference between pairs of matrices, divided by the number of cells in either matrix. Mean variation corresponds to the sum of absolute percentage variation of components in the matrix computed on the protected data with respect to components in the matrix computed on the original data, divided by the number of cells in either matrix. In the first two definitions, the distinction was on the type of distance used in the measure (square error vs. absolute error). In contrast, in the last definition, we have a relative error. Because of that, a change of the measure scale of the variables does not change the outcome of the measure.

The following definition formalizes the information loss measures in [2] for numerical microdata files.

Definition 7.2 Let X be a numerical microdata file and $X' = \rho(X)$ the protected version of the same file, let V and V' be the covariance matrices of X and X', let R and R' be the correlation matrices of X and X', let RF and RF' be the correlation matrices between the p variables and the p factors obtained through principal components analysis from X and X', let C and C' be the vector of commonalities[2] for X and X', let F and F' be the factor score coefficient matrices[3] for X and X'.

Let us now define three divergence functions to compare two matrices.

- $divergence_{MSE}(M, M') = \frac{\sum_{ij}(M_{ij}-M'_{ij})^2}{c(M)}$ (mean square error)
- $divergence_{MAE}(M, M') = \frac{\sum_{ij}|M_{ij}-M'_{ij}|}{c(M)}$ (mean absolute error)
- $divergence_{MRE}(M, M') = \frac{\sum_{ij}\frac{|M_{ij}-M'_{ij}|}{|M_{ij}|}}{c(M)}$ (mean relative error)

where $c(M)$ is the number of elements in the matrix. For example, for X we have $c(X) = n \cdot p$ where n is the number of records and p the number of attributes, while for R we have $c(R) = p \cdot p$ as M is a square matrix with as many rows as attributes we have in X.

Then, the following information loss measures are defined.

- $IL_{Id}(X, X') = divergence_{MSE}(X, X')$
- $IL_{Cov}(V(X), V(X')) = divergence_{MSE}(V, V')$
- $IL_{Corr}(R(X), R(X')) = divergence_{MSE}(R, R')$
- $IL_{CorrPCA}(RF(X), RF(X')) = divergence_{MSE}(RF, RF')$
- $IL_{CommPCA}(C(X), C(X')) = divergence_{MSE}(C, C')$
- $IL_{FsPCA}(F(X), F(X')) = divergence_{MSE}(F, F')$

Similar expressions can be defined with the other two divergence functions introduced before $divergence_{MAE}$ and $divergence_{MRE}$.

The expressions for divergence in the previous definition lead to unbounded information loss measures. Reference [21] considered this problem for the mean relative

---

[2]Commonality is the percentage of each variable that is explained by a principal component.
[3]The factor scores stand for the factors that should multiply each variable in X to obtain its projection on each principal component.

error of X and X' and proposed to use the following expression that is more stable when the original values are close to zero:

$$IL'_{Id}(X, X') = divergence'_{MRE}(X, X') = \frac{\sum_{ij} \frac{|x_{ij} - x'_{ij}|}{\sqrt{2}S_j}}{n \cdot p}$$

where S_j is the standard deviation of the jth attribute.

Reference [22] also discussed the unbounded measures and proposed to settle a predefined maximum value of error.

Reference [23] introduced probabilistic information loss measures to avoid predefined values. They assumed that X' is a sample from the population. Then, it considers the discrepancy between a population parameter θ on X and a sample statistic Θ on X'. Let $\hat{\Theta}$ be the value of this statistic for a specific sample. Then, the standardized sample discrepancy corresponds to

$$Z = \frac{\hat{\Theta} - \theta}{\sqrt{Var(\hat{\Theta})}}$$

This discrepancy is assumed to follow a $N(0, 1)$ (see [23] for details). Accordingly, the probabilistic information loss measure for $\hat{\Theta}$ was defined as follows:

$$pil(\hat{\Theta}) := 2 \cdot P\left(0 \leq Z \leq \frac{\hat{\theta} - \theta}{\sqrt{Var(\hat{\Theta})}}\right) \tag{7.2}$$

So, from our perspective, we have that information loss used the following expression for the divergence.

$$divergence(\theta, \hat{\Theta}) = 2 \cdot P\left(0 \leq Z \leq \frac{\hat{\theta} - \theta}{\sqrt{Var(\hat{\Theta})}}\right).$$

An alternative to these measures is to compare probability distributions. That is, consider that $f(X)$ corresponds to a probability distribution on some variables of X and then use a distance on probability distributions as the function *divergence*. [24] use the expected absolute difference, others (see e.g. [25]) use the Hellinger distance. Given a data set X, a probability distribution can be derived assuming a parametric model. Alternatively, a probability distribution is derived by means of a discretization of the domain of X and counting the number of records in each region. In this latter case, the more intervals are generated in the discretization, the more sensitive is the information loss to noise.

Finally, it is worth to mention that for microaggregation methods it is usual to use SSE/SST as a measure of information loss. See e.g. [26–31]. Recall from Eq. 6.2 that SSE is defined as

$$SSE(X, \chi, p) = \sum_{i=1}^{c} \sum_{x \in X} \chi_i(x)(d(x, p_i))^2$$

for a set of records X, a partition on X represented by χ, and cluster centers p. Note that the partition χ divides the set X into c clusters, and it is represented with the characteristic function χ_i where $\chi_i(x) = 1$ if the record x is assigned to the ith cluster and $\chi_i(x) = 0$ if not. Each part of the partition has a cluster center: p_i is the cluster center of the ith cluster. While SSE is the overall distance of elements x to the corresponding cluster centers, SST is the distance of elements to the mean of X, or, in other words, to a partition that contains a single cluster with all the elements. Formally,

$$SST(X) = \sum_{x \in X} (d(x, \bar{X}))^2$$

Note that as SST is constant for a given set X, optimization of SSE/SST is only about the optimization of SSE.

7.3.2 Categorical Data

In the case of categorical data, [1] uses a direct comparison of the categorical data. This is based on distances as the ones introduced in Sect. 5.4.7 (Definition 5.3). Reference [32] used for categorical values with an underlying hierarchical ontology the distance given in Definition 5.4.

An alternative to the comparison of records is the comparison of contingency tables. They have been used in a few works (see e.g. [1,33]) comparing tables built up to a given dimension. This is defined in terms of a divergence measure between contingency tables and using f as the function to build the contingency table. Reference [1] used the absolute distance for the comparison of the tables, while [33] uses the Hellinger distance[4] and a difference between entropies computed from the contingency tables. Note that these definitions are a categorical counterpart of comparing probability distributions for numerical data (discussed above).

Two other definitions based on entropy use the probability of replacing a category by another. We define them below.

Definition 7.3 Let X and X' be the original and protected files, let V be a categorical variable with values $1, \ldots, n$. Let us use $V = i$ and $V' = j$ to represent that X takes

---

[4]Reference [34] has a similar use of the Hellinger distance for comparing tables, but for tabular data protection.

value i in variable i and that X' takes value j for the same variable. Let $P_{V,V'} = \{p(V' = j|V = i)\}$ be the probability that value i has been replaced by value j. Then, the condicional uncertainty of V given $V' = j$ is defined in terms of the entropy of the distribution given j as follows

$$H(V|V' = j) = -\sum_{i=1}^{n} p(V = i|V' = j) \log p(V = i|V' = j).$$

Note that $p(V = i|V' = j)$ is computed from $P_{V,V'}$ using the Bayes' expression.

This definition was introduced in [35] for the PRAM method. Note in this case that $P_{V,V'}$ is the Markov matrix that PRAM requires. For other methods, we can estimate $P_{V,V'}$ from both X and X'.

Using the definition above, the entropy-based information loss measure can be defined as follows.

Definition 7.4 Let X, X', V, V', and $P_{V,V'}$ as above. Then, the entropy-based information loss (EBIL) is defined by

$$EBIL(P_{V,V'}, X') = \sum_{x' \in X'} H(V|V' = V(x')).$$

In [36] (Proposition 5) we gave an expression for the expected EBIL in the case of using a PRAM Markov matrix $P_{V,V'}$. It is the following one

$$-\sum_{i} n_i \sum_{j} p(V' = j|V = i) \sum_{k} p(V = k|V' = j) \log P(V = k|V' = j) \quad (7.3)$$

where n_i is the number of records in the file X that has value i. I.e,

$$n_i = |\{x \in X|V(x) = i\}|.$$

In EBIL, we have that the for a record x the computation of $H(V|V' = j)$ only takes into account the protected value but not the original one. Nevertheless, when we have a recoding of two values into a new one, information loss should depen on the proportion of these values in the file (or in the population). This is illustrated in the following example from [1,37].

Example 7.3 Let X, X', V and V' as above. Let V correspond to the variable town. The masking process for V is to replace location by states. Therefore, locations like New York City and Albany will be recoded into NY (New York State).

Then, all locations in NY will have the same entropy measure according to Definition 7.3.

According to the U.S.Census Bureau's American FactFinder (U.S. Census Bureau 2010) the population of New York State in 2010 was 19,378,102, the population of New York City was 8,175,133, and the population of Albany was 97,856. Thus, the conditional probabilities for NY are:

$$P(V =' Albany'|V = NY) = \qquad 97,856/19,378,102 = 0.005$$
$$P(V =' New\ York\ City'|V = NY) = \qquad 8,175,133/19,378,102 = 0.421$$

The entropy for recoding all records in the file which belong to NY requires the computation of the following summatory, which goes over the 932 towns and 62 cities in which NY is divided.

$$H(V|V' = NY) = - \sum_{i=1}^{n} p(V = i|V' = NY) \log p(V = i|V' = NY) = 1.69.$$

This measure is used for all records assigned as NY, independently of the size of the town. E.g., to records with an original value equal to New York City (8,175,133 inhabitants), Albany (97,856 inhabitants), and Red House (38 inhabitants).

In order to have a measure that depends on the relative size of the original town, the following per-record measure was introduced.

Definition 7.5 Let X, X', V, and V' as above. Let us use also $V = i$ and $V' = j$ as above. Then, we define the per-record information loss measure as:

$$PRIL(P_{V,V'}, i, j) = - \log p(V = i|V' = j).$$

Note that in this case,

$$PRIL(P_{V,V'}, New\ York\ City, NY) = - \log 0.421 = 0.8651$$
$$PRIL(P_{V,V'}, Albany, NY) = - \log 0.005 = 5.298$$

and, thus, the information loss of recoding someone in New York City is smaller than recoding someone in Albany. Some properties have been proven for PRIL. They are sumarized below.

Proposition 7.1 *[36] Let X, and V as above. Let $P_{V,V'}$ represent the Markov matrix for PRAM to be used to build X'. Then, the following holds.*

- *For a record in X with value i, its expected PRIL is:*

$$-\sum_j p(V' = j | V = i) \log p(V = i | V' = j).$$

- *For the whole data set X, the expected PRIL is*

$$\sum_i n_i e_i^T \hat{p} \, p(V' | V = i). \qquad (7.4)$$

  where e_i is the unit vector with one for the ith position and 0 otherwise, e_i^T its transpose, n_i is the number of records in the file X that have value i, and $\hat{p}$ is the matrix defined by $\{-\log p(V = i | V' = j)\}_{ij}$.
- *Let I_j be the values i such that $P(V' = j | V = i) \neq 0$. Then, expected EBIL equals expected PRIL when for all j, $P(V = i_1 | V' = j) = P(V = i_2 | V' = j)$ for all $i_1, i_2 \in I_j$.*

The last condition describes when the two measures EBIL and PRIL lead to the same results for a pair of files X and X'.

7.4   Specific Information Loss

In this section we review information loss focusing on the case of classification, regression, and clustering.

7.4.1   Classification-Based Information Loss

Classification is one of the most common data uses in machine learning. This is to extract a model for a categorical variable from the data (see Sect. 2.2.1). The (standard) goal of applying machine learning to build classifiers is that the model obtained is as accurate as possible. This is also the case when data is masked.

In this type of data use, research usually compares the results of different machine learning models using tables and figures. Authors compare the performance of models built using the original data and the ones using the masked data. Most of the literature uses accuracy [16,18,38,39], and others have used the area under the curve (AUC) [39].

Some experiments on building classifiers with masked data show that the performance of these classifiers is not always worse than the performance of the classifiers built using the original data. There are cases in which the performance is even improved. Reference [38] reports that "in many cases, the classification accuracy improves because of the noise reduction effects of the condensation process".

The same was concluded in [40] for recommender systems: "we observe that the prediction accuracy of recommendations based on anonymized ratings can be better than those based on non-anonymized ratings in some settings". The results in [16,39] are also relevant for this discussion. The reason seems to be that methods as microaggregation can be considered as methods for noise/dimensionality reduction. When the number of records is large and they are noisy, reducing the number of these records by means of averaging some of them has positive effects. In addition, when machine learning methods are resistant to errors, some data perturbation does not reduce dramatically the accuracy of the model.

An alternative to accuracy is to consider the similarities between the models themselves. I.e., if we obtain decision trees from X and X', we may want that the two trees are the same (or as similar as possible). Not only that they classify new data in the same way. This approach is discussed in [41] in the context of the definition of integral privacy. Note however, that for some classification models this approach is rather unfeasible as small variations of the data change the model. Observe the case of k-nearest neighbor. Model comparison has been used in regression problems as we discuss below.

7.4.2 Regression-Based Information Loss

Literature on statistical disclosure control considers regression and logistic regression as a usual data use. It is also present as a data use in some papers from the data mining community. Evaluation of regression can follow the pattern of classification. We can evaluate the models themselves or the performance of the model.

For example, [10] compares estimates of the regression models, while [7] compares the predictions of different models (sum of squares error). Reference [33] considers a measure based on the optimal log-linear model. Some of these comparisons are given graphically, plotting a 2D graph with one prediction in one dimension and the other prediction in the other axis.

In addition to the analysis of the model, it is quite common to consider in generic information loss measures correlation coefficients (see e.g. [7,9–12]).

7.4.3 Clustering-Based Information Loss

Clustering is a typical data analysis in both the statistical and machine learning communities. Because of that, masking methods have been evaluated and compared against its performance with respect to clustering algorithms. In order to do so, we define the two components (i.e., the exact analysis f and the divergence in Eq. 7.1).

The function is one clustering algorithm with its parameters (e.g., the number of clusters). The divergence is to select a function to compare the results of the clustering.

Although there are several types of cluster structures (i.e., clusters, fuzzy clusters, dendrograms), analysis of data masking algorithms has been focused on crisp (mainly from k-means) and fuzzy clusters.

In order to compare crisp clusters we can use distances and indices to compare partitions. The Rand, Jaccard, and Wallace indices, the adjusted Rand index, and the Mántaras distance (see Sect. 2.3.1) have been used for this purpose. See e.g. [42,43]. In [15], the F-measure is used to compare partitions.

In Sect. 2.3.1 we also discussed how to compare fuzzy clusters. These approaches are suitable for measuring information loss when fuzzy clustering is used. Reference [44] uses a distance on α-cuts for this purpose.

Clustering is used extensively within the machine learning community for all types of data (including standard databases, time series, access and search logs, documents, social networks, etc.). In all these cases we can follow the same approach. That is, we can compare partitions (of records, time series, etc.). For example, [43] compares clusters of query logs.

7.5 Information Loss and Big Data

Evaluation of information loss for big data follows the same approach formalized in Definition 7.1. That is, we need to evaluate the divergence between analysis for the original data set and the ones for the protected data set. The difficulties we found are due to the nature of the analysis we apply to big data. We can discuss them in relation to the three categories we have considered in previous chapters for big data: (i) large volumes, (ii) streaming data, and (iii) dynamic data.

- **(i) Large volumes**. The analysis of standard databases of huge dimensions and the analysis of typically large datasets (e.g., social networks, location data) has been developed in the last years within the field of data mining. There are effective algorithms for clustering and classification, and also for computing specific indices for some particular types of data.

   Research challenges focus on how to compare efficiently the summarizations of the data when these summaries are also of large dimension. In addition, some of the algorithms are not deterministic. This means that even with the same data, different outcomes can be obtained. Because of that effective computation of information loss is difficult. In short, it is difficult to know in what extent divergence in the analysis is due to divergences on the data or to variability on the results.

- **(ii) Streaming data**. The research problems of large volumes also apply here. In addition, we have that comparison of results need to be done at run time. This implies that we do not have the full picture of $f(X)$ and $f(X')$ at any instant but only part of this function.

- **(iii) Dynamic data**. Information loss needs to consider that at a given point we have a set of databases $X_1, X_2, \ldots, X_n$ together with their masked counterparts $X'_1, X'_2, \ldots, X'_n$. Aggregation of individual divergences

$$d_i = divergence(f(X_i), f(X'_i))$$

using an aggregation function $\mathbb{C}$ as e.g. the arithmetic mean or a weighted mean is the most straightforward way to deal with this problem. An alternative is to consider analysis of several releases. That is

$$IL = divergence(f(X_1, X_2, \ldots, X_n), f(X'_1, X'_2, \ldots, X'_n)).$$

Up to our knowledge, no much research has been done in this area.

We have discussed extensively that data perturbation causes that analysis on the original data and on the protected data differ. This implies some information loss. Nevertheless, it is important to underline that for some data uses, the noise introduced with a data masking procedure still permits us to obtain acceptable results. Naturally this depends on the data and the data use. We have reported in Sect. 7.4.1 the case of building classifiers. Two notes can be made in this respect.

On the one hand, machine learning algorithms are being designed so that they are resistant to errors, and machine learning design strategies are developed to avoid overfitting. In this way, algorithms can deal with some noise added to the data either accidentally or on purpose.

On the other hand, some masking mechanisms can be seen (or can be developed) as privacy-providing noise reduction methods. It is usual in machine learning to preprocess the data to improve their quality. Variable selection and dimensionality reduction algorithms are used. Microaggregation and other methods to achieve *k*-anonymity are examples of masking methods that can be seen from this perspective. The process of replacing a set of a few similar data elements by their mean can be seen as a way of consolidating the data and reducing the error.

References

1. Domingo-Ferrer, J., Torra, V.: Disclosure control methods and information loss for microdata. In: Doyle, P., Lane, J.I., Theeuwes, J.J.M., Zayatz, L. (eds.) Confidentiality, Disclosure, and Data Access: Theory and Practical Applications for Statistical Agencies, North-Holland, pp. 91–110 (2001)
2. Domingo-Ferrer, J., Torra, V.: A quantitative comparison of disclosure control methods for microdata. In: Doyle, P., Lane, J.I., Theeuwes, J.J.M., Zayatz, L. (eds.) Confidentiality, Disclosure and Data Access: Theory and Practical Applications for Statistical Agencies, North-Holland, pp. 111–134 (2001)

3. Rebollo-Monedero, D., Forné, J., Soriano, M.: An algorithm for k-anonymous microaggregation and clustering inspired by the design of distortion-optimized quantizers. Data Knowl. Eng. **70**(10), 892–921 (2011)

4. Rebollo-Monedero, D., Forné, J., Pallarés, E., Parra-Arnau, J.: A modification of the Lloyd algorithm for k-anonymous quantization. Inf. Sci. **222**, 185–202 (2013)

5. Drechsler, J., Reiter, J.P.: An empirical evaluation of easily implemented, nonparametric methods for generating synthetic datasets. Comput. Stat. Data Anal. **55**, 3232–3243 (2011)

6. Liu, L., Wang, J., Zhang, J.: Wavelet-based data perturbation for simultaneous privacy-preserving and statistics-preserving. In: IEEE ICDM Workshops (2008)

7. Muralidhar, K., Sarathy, R.: An enhanced data perturbation approach for small data sets. Decis. Sci. **36**(3), 513–529 (2005)

8. Kim, J., Winkler, W.: Multiplicative noise for masking continuous data, U.S. Bureau of the Census, RR2003/01 (2003)

9. Carlson, M., Salabasis, M.: A data swapping technique using ranks: a method for disclosure control. Res. Off. Stat. **5**(2), 35–64 (2002)

10. Raghunathan, T.E., Reiter, J.P., Rubin, D.B.: Multiple imputation for statistical disclosure limitation. J. Off. Stat. **19**(1), 1–16 (2003)

11. Reiter, J.P., Drechsler, J.: Releasing multiply-imputed synthetic data generated in two stages to protect confidentiality. Stat. Sinica **20**, 405–421 (2010)

12. Drechsler, J., Bender, S., Rässler, S.: Comparing fully and partially synthetic datasets for statistical disclosure control in the German IAB Establishment Panel. Trans. Data Priv. **1**, 105–130 (2008)

13. Reiss, S.P.: Practical data-swapping: the first steps. ACM Trans. Dataase Syst. **9**(1), 20–37 (1984)

14. Liu, K., Kargupta, H., Ryan, J.: Random projection based multiplicative data perturbation for privacy preserving data mining. IEEE Trans. Knowl. Data Eng. **18**(1), 92–106 (2006)

15. Hajian, S., Azgomi, M.A.: A privacy preserving clustering technique using Haar wavelet transform and scaling data perturbation. IEEE (2008)

16. Bapna, S., Gangopadhyay, A.: A wavelet-based approach to preserve privacy for classification mining. Decis. Sci. **37**(4), 623–642 (2006)

17. Mukherjee, S., Chen, Z., Gangopadhyay, A.: A privacy-preserving technique for Euclidean distance-based mining algorithms using Fourier-related transforms. VLDB J. **15**, 293–315 (2006)

18. Agrawal, R., Srikant, R.: Privacy preserving data mining. In: Proceedings of the ACM SIGMOD Conference on Management of Data, pp. 439–450 (2000)

19. Domingo-Ferrer, J., Mateo-Sanz, J. M., Torra, V.: Comparing SDC methods for microdata on the basis of information loss and disclosure risk. In: Pre-proceedings of ETK-NTTS 2001, vol. 2, pp. 807–826. Eurostat (2001)

20. Domingo-Ferrer, J., González-Nicolás, U.: Hybrid microdata using microaggregation. Inf. Sci. **180**, 2834–2844 (2010)

21. Yancey, W.E., Winkler, W.E., Creecy, R.H.: Disclosure risk assessment in perturbative microdata protection. In: Domingo-Ferrer, J. (ed.) Inference Control in Statistical Databases. LNCS, vol. 2316, pp. 135–152 (2002)

22. Trottini, M.: Decision models for data disclosure limitation, Ph.D. Dissertation, Carnegie Mellon University (2003)

23. Mateo-Sanz, J.M., Domingo-Ferrer, J., Sebé, F.: Probabilistic information loss measures in confidentiality protection of continuous microdata. Data Min. Knowl. Disc. **11**(2), 181–193 (2005)

24. Agrawal, D., Aggarwal, C.C.: On the design and quantification of privacy preserving data mining algorithms. In: Proceedings of the PODS 2001, pp. 247–255 (2001)

25. Torra, V., Carlson, M.: On the Hellinger distance for measuring information loss in microdata, UNECE/Eurostat Work Session on Statistical Confidentiality, 8th Work Session 2013, Ottawa, Canada (2013)
26. Domingo-Ferrer, J., Mateo-Sanz, J.M.: Practical data-oriented microaggregation for statistical disclosure control. IEEE Trans. Knowl. Data Eng. **14**(1), 189–201 (2002)
27. Laszlo, M., Mukherjee, S.: Minimum spanning tree partitioning algorithm for microaggregation. IEEE Trans. Knowl. Data Eng. **17**(7), 902–911 (2005)
28. Chang, C.-C., Li, Y.-C., Huang, W.-H.: TFRP: an efficient microaggregation algorithm for statistical disclosure control. J. Syst. Softw. **80**, 1866–1878 (2007)
29. Panagiotakis, C., Tziritas, G.: Successive group selection for microaggregation. IEEE Trans. Knowl. Data Eng. **25**(5), 1191–1195 (2013)
30. Laszlo, M., Mukherjee, S.: Iterated local search for microaggregation. J. Syst. Soft. **100**, 15–26 (2015)
31. Cheng, L., Cheng, S., Jiang, F.: ADKAM: A-diversity k-anonymity model via microaggregation. In: Proceedings of the ISPEC 2015. LNCS, vol. 9065, pp. 533–547 (2015)
32. Salari, M., Jalili, S., Mortazavi, R.: TBM, a transformation based method for microaggregation of large volume mixed data. Data Min. Knowl. Discov. (2016, in press). doi:10.1007/s10618-016-0457-y.
33. Gomatam, S., Karr, A.F., Sanil, A.P.: Data swapping as a decision problem. J. Off. Stat. **21**(4), 635–655 (2005)
34. Shlomo, N., Antal, L., Elliot, M.: Measuring disclosure risk and data utility for flexible table generators. J. Off. Stat. **31**(2), 305–324 (2015)
35. Willenborg, L., de Waal, T.: Elements of Statistical Disclosure Control. Lecture Notes in Statistics. Springer, New York (2001)
36. Torra, V.: Progress report on record linkage for risk assessment. DwB project, Deliverable 11.3 (2014)
37. Torra, V.: On information loss measures for categorical data, Report 3, Ottilie Project (2000)
38. Aggarwal, C.C., Yu, P.S.: A condensation approach to privacy preserving data mining. In: Proceedings of the EDBT, pp. 183–199 (2004)
39. Herranz, J., Matwin, S., Nin, J., Torra, V.: Classifying data from protected statistical datasets. Comput. Secur. **29**, 875–890 (2010)
40. Sakuma, J.: Recommendation based on k-anonymized ratings. Arxiv (2017)
41. Torra, V., Navarro-Arribas, G.: Integral privacy. In: Proceedings of the CANS 2016. LNCS, vol. 10052, pp. 661–669 (2016)
42. Ladra, S., Torra, V.: On the comparison of generic information loss measures and cluster-specific ones. Int. J. Unc. Fuzz. Knowl. Based Syst. **16**(1), 107–120 (2008)
43. Batet, M., Erola, A., Sánchez, D., Castellà-Roca, J.: Utility preserving query log anonymization via semantic microaggregation. Inf. Sci. **242**, 49–63 (2013)
44. Torra, V.: On the definition of cluster-specific information loss measures. In: Solanas, A., Martínez-Ballesté, A. (eds.) Advances in Artificial Intelligence for Privacy Protection and Security, pp. 145–163. World Scientific (2009)

Selection of Masking Methods

8

> No hauria sabut explicar la felicitat que sentia
> abocada al balcó d'aquella casa, menjant borregos
> i una presa de xocolata darrera de l'altra (...)
>
> M. Rodoreda, Mirall trencat, 1974 [1]

We have discussed in Chap. 5 (Sect. 5.1.1) that selection of a masking algorithm (and its parameters) can be seen as an optimization problem. When we consider a Boolean privacy model, we need to select the method that minimizes information loss. In contrast, when we consider a measurable privacy model, we need to select the method that is the best in a multicriteria optimization problem. This is so because we need the method to be the best with respect both the information loss and the disclosure risk. In this latter case, if we want to compare different alternatives (i.e., different masking methods and parameters), we can either aggregate the two measures or visualize the outcome.

8.1 Aggregation: A Score

The simplest way to give a trade-off between information loss and disclosure risk for a given masking method m with parameters p and given a data set X is to compute their average.

$$Score(m, p, X) = \frac{IL(m, p, X) + DR(m, p, X)}{2}.$$

© Springer International Publishing AG 2017
V. Torra, *Data Privacy: Foundations, New Developments
and the Big Data Challenge*, Studies in Big Data 28,
DOI 10.1007/978-3-319-57358-8_8

where *IL* corresponds to the information loss and *DR* to the disclosure risk. Different measures for *IL* and *DR* lead to different scores. This score was first used in [2] to compare a plethora of methods with different parameters. They defined *IL* and *DR* in terms of an aggregation of a few information loss and disclosure risk measures.

This definition presumes that both *IL* and *DR* are in the same scale, say [0, 100], and implies that an increment of 1 in *IL* can be compensated with a decrement of 1 in *DR*. Because of that, the following three cases (i.e., three straightforward *masking* methods) are considered as equivalent:

- **The identity method**. A method that publishes the original data without perturbation leads to the minimal information loss ($IL = 0$), the maximum risk ($DR = 100$) and, thus, a score of 50.
- **The random method**. A method that generates a completely random file, that has no resemblance with the original file and thus it is useless for data analysis, has the maximal loss ($IL = 100$), the minimum risk ($DR = 0$) and, thus, a score of 50.
- **The 50-points method**. Any method that manipulates the data so that $IL = DR = 50$ will also lead to a score of 50.

These considerations also imply that any approach leading to a file with a score larger than 50 is not worth to consider because this trade-off is worse than just generating completely random data, or publishing the original file without perturbation.

Other types of aggregation functions can be used to change the level curves of the score. That is, to give larger importance to one of the components or not allowing compensation between DR and IL. For example, we can use the maximum (as in [3]) defining

$$Score(m, p, X) = \max(IL(m, p, X), DR(m, p, X))$$

that will favor methods that are good for both *IL* and *DR*. I.e., a method with $IL = DR = 30$ is better rated that one with $IL = 20$ and $DR = 40$, and thus, no so much compensation is allowed. Discussion on aggregation functions and their properties can be found in [4–6].

8.2   Visualization: R-U Maps

When we have the two dimensions of risk and information loss, we can visualize the performance of a method with a two dimensional graph. The R-U maps, first proposed in [7,8] are such graphical representation. They are the risk-utility maps. Similarly, we can represent IL-DR maps for information loss and disclosure risk. In an IL-DR map the best methods are the ones that are nearer to (0, 0). Following the discussion in the previous section all methods above the line (0, 100)–(100, 0) are not worth to consider. They are the ones with a score larger than 50.

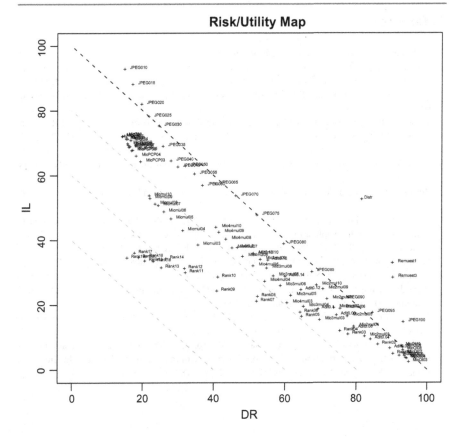

Fig. 8.1 R-U Maps for some protection methods (from [9]). Information loss computed with PIL

Figure 8.1 is an example of a IL-DR map from [9] that compares the measures for a few masking methods and parameterizations.

8.3   Optimization and Postmasking

When we consider methods as an optimization problem, we can consider the combination of a masking method and a post-masking method. The post-masking is to improve the value of the measures further modifying the data.

This can be done by means of e.g. a kind of normalization (to achieve appropriate means and variances) or applying adhoc methods (as e.g. genetic algorithms just to increase the overall measure). See e.g. [10] for an example of a post-masking approach. Methods described in Sect. 6.5 that are designed to solve the violation of constraints can be seen from this perspective.

References

1. Rodoreda, M.: Mirall trencat, Club editor (1974)
2. Domingo-Ferrer, J., Torra, V.: A quantitative comparison of disclosure control methods for microdata. In: Doyle, P., Lane, J.I., Theeuwes, J.J.M., Zayatz, L. (eds.) Confidentiality, Disclosure and Data Access: Theory and Practical Applications for Statistical Agencies, pp. 111–134. North-Holland (2001)
3. Marés, J., Torra, V., Shlomo, N.: Optimisation-based study of data privacy by using PRAM. In: Navarro-Arribas, G., Torra, V. (eds.) Advanced Research in Data Privacy, pp. 83–108. Springer (2015)
4. Torra, V., Narukawa, Y.: Modeling Decisions: Information Fusion and Aggregation Operators. Springer (2007)
5. Beliakov, G., Pradera, A., Calvo, T.: Aggregation Functions: A Guide For Practitioners. Springer (2007)
6. Grabisch, M., Marichal, J.-L., Mesiar, R., Pap, E.: Aggregation Functions. Encyclopedia of Mathematics and its Applications, no. 127. Cambridge University Press (2009)
7. Duncan, G.T., Keller-McNulty, S.A., Stokes, S.L.: Disclosure risk vs. data utility: the R-U confidentiality map. Technical report 121, National Institute of Statistical Sciences (2001)
8. Duncan, G.T., Keller-McNulty, S.A., Stokes, S.L.: Database security and confidentiality: examining disclosure risk vs. data utility through the R-U confidentiality map. Technical report 142, National Institute of Statistical Sciences (2001)
9. Torra, V.: Privacy in data mining. In: Maimon, O., Rokach, L. (eds.) Data Mining and Knowledge Discovery Handbook, pp. 687–716. Springer (2010)
10. Sebe, F., Domingo-Ferrer, J., Mateo-Sanz, J.M., Torra, V.: Post-masking optimization of the tradeoff between information loss and disclosure risk in masked microdata sets. In: Inference Control in Statistical Databases. LNCS, vol. 2316, pp. 187–196 (2002)

Conclusions

9

Jodi lea buoret go oru
Sami proverb [1]

This book has given an introduction to data privacy. We have presented the main areas and some of the methods and tools to ensure privacy and avoid disclosure.

We have presented different privacy models and disclosure risk measures. It is important to underline that different privacy models focus on different aspects of privacy. We have seen, for example, attribute and identity disclosure, and then mechanisms to protect these different types of disclosure. The methods that are suitable for one type of model may be unsuitable for the other. The selection of our privacy model will depend on the context (e.g., data and privacy threats).

For example, we have seen that methods ensuring k-anonymity may be unsuitable for attribute disclosure. l-Diversity tries to consider both types of disclosures, at the cost of a larger information loss.

We have seen that differential privacy considers privacy from a different perspective. The goal is to avoid that intruders learn whether a record is in the database. With this purpose in mind, it masks the result of a query usually adding some noise. Because of that, when the query is e.g. the mean of one of the variables of the file, it is possible that the result diverges (even largely) from the correct mean. On the contrary, masking methods or methods that are compliant with k-anonymity can produce a masked data set with a mean that is similar or equal to the original mean. The latter methods can ensure that the probability of linking a given record to the masked file is lower than a certain threshold, but they are not compliant with differential privacy definition.

Finally, computation-driven (cryptographic) approaches usually achieve 100% exact output, 100% privacy but they do not provide flexibility in the computation

© Springer International Publishing AG 2017
V. Torra, *Data Privacy: Foundations, New Developments and the Big Data Challenge*, Studies in Big Data 28, DOI 10.1007/978-3-319-57358-8_9

and they are not compliant with differential privacy (unless they are explicitly defined to do so and then they will not be exact).

In Chap. 1 we have presented the principles of privacy by design. We have seen in this book tools that permit to develop systems that help us to follow these principles. For example, the first principle is that privacy should be proactive. The guidelines discussed in Sect. 5.9.4 are of relevance here. User privacy and decentralized anonymity should be put in practice, and databases should be anonymized in origin. Controlled linkability should permit us to link databases even when they are already protected and linked databases should avoid disclosure. The implementation of these guidelines follow some of the other principles of privacy by design. As we have seen in this book methods exist to provide solutions for some of these technological problems, but further research is needed to solve the other.

We have also advocated in this book the need for transparency. Transparency is also one of the principles of privacy by design. Data protection methods should be resistant to transparency attacks. We have discussed this issue extensively in Chap. 6 and we have seen that there are masking methods that have been designed to be resistant to transparency attacks. Unfortunately, however, new effective transparency attacks may be further designed for masking methods unless privacy guarantees have not been formally proven. The study of attacks to masking methods and, specially, transparency attacks is essential in order to have accurate assessments of disclosure risk.

Another point is that data privacy should be knowledge intensive. In the era of big data, when data of all kinds are integrated at large, and when ontologies and natural-language processing technologies are used for all type of applications, data privacy needs to take advantage of these technologies to have accurate estimations of the risk, and provide effective solutions for ensuring privacy. We have mentioned this issue in Sect. 3.1.4.

A final consideration. Data privacy is a complex area that is influenced by variables in different dimensions (technological, sociological, psychological, legal, economical, political, educational, ethical, etc.). In this book we have focused on the technological aspects. To ensure privacy, the development of the technological dimension is not enough, these other dimensions need to be taken into account.

Reference

1. Gaski, H.: Time Is a Ship That Never Casts Anchor, ČálliidLágádus (2006)

Index

A

Access logs
  information loss, 250
  k-anonymity, 225
  microaggregation, 212
Accountability, 14
Accuracy, 27
Actees, 7
Actor, 7
Additive noise, 212, 217
  big data, 213
  correlated noise, 212
  semantics, 213
  uncorrelated noise, 212
  with constraints, 229
Adjusted Rand index, 43, 250
Adversarial search engine, 99
Adversary, 7
Aggregation function, 131
Anomality detection, 182
Anonymity, 8, 9, 14
  cannot increase, 13
  definition, 8
Anonymity preserving pattern discovery, 71
Anonymization in origin, 182
Anonymization methods, 62
Anonymous communications, 2, 90
Anonymous database search, 97, 98
AOL case, 3, 15, 112
Apriori algorithm, 48
Arithmetic mean, 164
Assignment problem, 125
Association rule, 44
  apriori algorithm, 48
  confidence, 46, 72

  hiding, 71
  matching, 45
  result-driven, 71
  support, 45, 72
  support count, 45
Attack
  external, 77
  external knowledge, 177
  homogeneity, 177
  internal, 77
  internal with dominance, 77
Attacker, 7
Attribute disclosure, 9, 111, 112, 115, 120
  categorical, 116, 118
  definition, 10
  numerical, 115

B

Belief measures, 179
Bell number, 29
Big data, 65, 260
  additive noise, 213, 230
  disclosure risk, 180
  information loss, 250
  k-anonymity, 225, 230
  lossy compression, 218
  microaggregation, 211, 230
  multiplicative noise, 213
  PRAM, 216, 230
  rank swapping, 200, 230
Bigrams, 139
Bipartite graph, 174
Blocking, 135, 161
  cluster-based, 135
Blocking variables, 135

© Springer International Publishing AG 2017
V. Torra, *Data Privacy: Foundations, New Developments and the Big Data Challenge*, Studies in Big Data 28,
DOI 10.1007/978-3-319-57358-8

Printed in the United States
By Bookmasters